**Distribution Sampling for
Computer Simulation**

Distribution Sampling for Computer Simulation

T. G. Lewis
University of Southwestern
Louisiana

Lexington Books
D.C. Heath and Company
Lexington, Massachusetts
Toronto   London

Library of Congress Cataloging in Publication Data

Lewis, Theodore Gyle, 1941
    Distribution sampling for computer simulation.

    Includes bibliographical references.
    1. Electronic data processing—Sampling (Statistics) 2. Random number generators. 3. Electronic data processing—Distribution (Probability theory) 4. Digital computer simulation. I. Title.
QA276.7.L4   519.5   74-25058
ISBN 0-669-97139-1

Published simultaneously in Canada.

Printed in the United States of America.

International Standard Book Number: 0-669-97139-1

Library of Congress Catalog Card Number: 74-25058

Contents

To Carol

"Randomness is too unreliable!" she said,
(but it makes the simulation go).
"It's still unpredictable!" she demanded,
(that's why I make it pseudo).

List of Figures

Preface

This book on distribution sampling is divided into two parts. Part I is an exhaustive study of pseudorandom number generators. The generators are given in algorithmic form, and their n-space filling properties are emphasized.

The reader will notice that all modern pseudorandom number generators fall into basically two categories: congruential and shift register. In fact, they are the same. The congruential is based on a residue class: integers computed modulo some period, m. The shift register pseudorandom number generator is based on a residue class: polynomials computed modulo some primitive polynomial. The primitive polynomial is a member of Galois field of binary base $GF(2)$. Thus, the two types are quite analogous and very competitive in terms of computer implementation.

This study is clearly oriented toward computer implementation. That is, emphasis is on computer science aspects more than mathematical theory. The treatment does provide a definitive summary of mathematical theory to date. For example, wave properties and lattice phenomena are examined in a new light that gives the reader a unified treatment of these recent developments. The n-cube theorem provides a summary of the lattice structure and at the same time shows how it must occur.

The computer science emphasis is clear by the consideration given to implementation and the many computer programs given. The statistical tests of Chapter 4 are empirical. The author is firmly convinced that empirical tests are the only truly worthwhile and meaningful tests.

Much of the material is new. The generalized feedback shift register algorithm has been published in only a few journals and represents the "ultimate" in speed, generality, and length of period. The conditional bit sampling technique represents a step forward in generality and accuracy (see Part II).

Part II is a comprehensive study of traditional and recent tecnhiques in nonuniform distribution sampling. This is apparent by the chapter titles: Chapter 5 on inversion techniques, Chapter 6 on rejection techniques, and Chapter 7 on other techniques. The conditional bit sampling technique is new and perhaps somewhat untried. It does, however, represent a general technique and a new approach. Chapter 7 has some thought-provoking ideas for parallel processing machines.

The information in the Appendixes is perhaps of greatest interest to the practitioner. Here many tables, programs, and test results have been given. The programs are written mainly for IBM 360 computers, but should be easily translatable to other machines. The tables and test results are presented for the reader's interpretations.

Most important, the author owes thanks to Prof. W. H. Payne who patiently guided him during the research that led to the writing of this book. Dr. Payne stimulated many ideas and is the source of much of this work. Warren Camp contributed by running programs and finding errors before they became "bugs." Also to the manufacturers of the Calcomp Plotter, whose equipment generated many of the drawings used in this book. Mary Dieters is to be thanked for her typing, and the support given by the computer science departments at Washington State University, University of Missouri at Rolla, and the University of Southwestern Louisiana is appreciated.

**Distribution Sampling for
Computer Simulation**

**Part I
Pseudorandom Numbers**

Contemporary Pseudorandom Number Generators

1-1 Introduction

Computer simulation of complex problems is an increasingly common problem-solving approach, particularly when analytic solutions are difficult or impossible to obtain. Pseudorandom number generation (RNG) provides the foundation of all probabilistic and many deterministic simulation models [1]. The optimum pattern of stoplight settings in a traffic grid may be impossible to determine analytically, but simulation of such a network is a quick and simple way to determine such settings. Monte Carlo techniques use vast amounts of random numbers to compute deterministic quantities, e.g., the value of integrals. An RNG which, in fact, produces "random" numbers is essential.

A continuous, uniform random number ranges between zero and unity; that is, the random number is sampled from the uniform probability distribution function. A discrete, uniform random number to the base B takes on the values $0, 1, \ldots, B - 1$ with probability $1/B$, and each digit is independently selected. Tables of random digits were obtained by early investigators by using a spinning disc (1938), selective service lists (1942), least significant digits of logarithms (1947), special physical devices (1951), and lottery drawings (1952) [2].

In contrast to obtaining numbers from chance behavior of nature or probabilistic experiments, deterministic methods may be used in high-speed computers to generate pseudorandom numbers. However, deterministic methods raise the question of "randomness" and thus the prefix "pseudo" is appended to the name "random" to describe such sequences.

Lehmer's definition of a pseudorandom sequence is

. . . a vague notion embodying the idea of a sequence in which each term is unpredictable to the uninitiated and whose digits pass a certain number of tests . . . depending somewhat on the uses to which the sequence is to be put. [3]

This definition avoids philosophical questions of "true randomness," and it emphasizes the relationship between pseudorandom sequences and tests of randomness. Thus, a sequence is random relative to particular tests that are passed.

Speculation as to randomness of a sequence is based on empirical and/or analytic tests. Analytic tests on the full period of periodic sequences

3

are by no means sufficient. Empirical tests on fractional period sequences must be relied upon to establish randomness. Furthermore, even finite-state computers are capable of generating nonperiodic pseudorandom sequences which defy analytic examination. Pathria [4, 5] has empirically shown that the first 10,000 digits of pi and the first 2500 digits of e and $1/e$ are random relative to the classical tests of Kendall and Smith and Yule [6, 7]. Randomness tests, therefore, play an important role in RNG theory.

Digital computers have been equipped with special physical hardware devices which produce random numbers, but most random numbers are generated by software RNGs. Software RNGs are convenient to use, have repeatability, and constant mean and variance over time, but are low in speed/cost ratio compared with hardware devices [8]. The advantages of both hardware and software generators can be realized with micropro-gramming techniques.

RNGs which have been faithful for years may fail simple n-space tests for n greater than 20. Also, apparent one-dimensional problems may actually be n-dimensional problems in disguise. For example, computing expected run length in a sequence of random digits appears to be a one-dimensional problem, but actually corresponds with locating points in n-space. Multidimensional transformations from a group of pseudorandom numbers to a single variate may unravel the sequence revealing regularities unbecoming of truly random sequences.

Our purpose in this chapter and the next three chapters is to investigate current RNGs, to give new insight into their n-space behavior, and to present a completely general RNG that produces the same sequence of floating-point pseudorandom numbers for any binary computer. The precision of each number is determined by the number of bits in the floating-point mantissa. Such a RNG must also possess desirable n-space properties. A high speed/cost ratio is accomplished by implementation of the general RNG as a microprogrammed instruction in read-only memory.

This chapter describes the Lehmer and Tausworthe RNGs in current use. Both are based on residue arithmetic; the Lehmer operates on a field of integers and the Tausworthe over a polynomial field modulo 2.

In Chapter 2, the n-space behavior of Lehmer and Tausworthe RNGs is investigated; n-space sparseness is shown to occur for all periodic sequences. That is, nonuniformity increases as dimensionality increases. A Fourier analysis performed on a bank of Lehmer generators shows them to be potentially deficient in n-space also. Extended sequences having repeated values within a very long period are possible, and feedback shift register sequences offer the greatest hope for n-space improvement.

In Chapter 3, a completely generalized algorithm possessing desirable n-space properties is investigated. In addition, this new algorithm offers advantages of speed and portability. For example, a microprogram for the

Interdata computer yields a speed/cost ratio advantage over software implementations.

Finally, exhaustive testing of the Lehmer, the Tausworthe, and the generalized algorithms is reported in Chapter 4. Low- and high-dimensional tests are applied, but fail to give conclusive indications of n-space nonuniformity. The generalized algorithm, with 15-bits of accuracy, performs satisfactorily as compared with 31-bit accuracy of the others tested.

The Appendixes that follow Part II contain many valuable (IBM 360, and FORTRAN) computer programs for executing the RNGs described here. Also, a set of test programs is given that may be easily used as a test package.

1-2 Pseudorandom Number Generators

The Lehmer RNG [9] has gained wide acceptance as a means of arithmetically producing long period sequences which pass most randomness tests. A complete account of the Lehmer theory with history and bibliography is found in Hull and Dobell [10]. For binary computers, the following theorems apply.

HULL-DOBELL THEOREM (HD1). The sequence defined by the mixed congruential relation,

$$X_i \equiv aX_{i-1} + c \quad (\text{mod } m)$$

has full period, m, provided that

(1) c is odd

(2) $a \equiv 1 \quad (\text{mod } 4)$

HULL-DOBELL THEOREM (HD2). The sequence defined by the multiplicative relation

$$X_i \equiv aX_{i-1} \quad (\text{mod } m)$$

has a maximal period, $m - 1$, provided

(1) X_0 is relatively prime to m.

(2) a is a primitive root for p^k if p^k is a factor of m, p odd, and k as large as possible, or $p = 2$, $k = 1$ or $k = 2$.

(3) a belongs to 2^{k-2}, if 2^k is a factor of m with $k = 2$. The maximal period is the LCM of the periods $(p - 1)p^{k-1}$ or 2^{k-2}, WRT the prime power factors.

In practice, it is easy to satisfy (1), and (3) is satisfied by insisting that $a \equiv \pm 3 \pmod 8$. However, finding primitive roots is tedious [11]. A list of primitive roots for $m = 2^{31} - 1$ (32-bit machine) is given in Figure 1-1. These primitive roots have been tested for randomness (Yule, Kendall and Smith tests) in HD2 (Hull-Dobell Theorem 2) when used in batches of 10^i, $i = 2, 3,$ 4, 5, and $X_0 = 524287$.

Variations and extensions of the Lehmer generator have been made, but the simplicity of HD1 has made it by far the most popular RNG [12, 13, 14, 15]. Greenberger [16] suggests selecting multipliers, a, near $\sqrt{m}$ in order to minimize autocorrelation. Whittlesey [17] has performed empirical tests which confirm Greenberger's analysis for some but not all multipliers. Coveyou and MacPherson [18] present Fourier analysis of HD1 which shows disturbingly low "wave numbers" in n-space. Marsaglia [19, 20] proved the existence of regularities in HD1 which becomes significant for high-dimensional n-space and decreasing word size.

Tausworthe [21] demonstrated the use of shift register sequences for RNGs based on primitive trinomials over $GF(2)$ [22, 23, 24]. Primitive polynomials over a binary field are completely analogous with primitive roots of HD2 except that polynomial algebra is used in place of multiplication of integers of a residue class.

TAUSWORTHE THEOREM (FEEDBACK SHIFT REGISTERS, FSR).   Let $a = \{a_k\}$ be the sequence of 0's and 1's generated by the linear recursion relation

$$a_k = \sum_{i=1}^{n} c_i a_{k-i} \qquad (\text{mod } 2)$$

for any set $c = \{c_i\}$ having a value of 0 or of 1 and $c_n = 1$. If

$$f(x) = \sum_{i=1}^{n} c_i x^i$$

is a primitive nth-degree polynomial over $GF(2)$, the sequence, a, is a maximal-length linearly recurring sequence modulo 2 with period $2^n - 1$. Furthermore, if

$$y_k = \sum_{t=1}^{L} 2^{-t} a_{qj+r-t}$$

for some $0 \le r \le 2^n - 1$, $L \le n$, $(q, 2^n - 1) = 1$, and $q \ge L$, then $y = \{y_k\}$ is distributed uniformly with

(1) $u \approx 0$, $\sigma^2 \approx 1/12$

(2) Autocorrelation, $R \approx 0$, var $(R) \approx 1/N$; N = sample size

(3) The relative number of times (frequency) that $(y_k, y_{k-l_2}, \ldots, y_{k-l_s})$ falls in an interval $1/2^{d_1} \cdot 1/2^{d_2} \ldots 1/2^{d_s}$ in the unit s-cube is

b	c						
7	5	43	71				
11	13	19	37	79			
14	17	29	37	41	47	79	
22	17	19	37	41	59		
28	19	41	73				
31	19	47	59	67			
39	13	23	41	53	59	61	67
44	17	37	41	67	73		
45	5	17	23				
51	13	17					
56	37	43	59	67			
57	23	29	37	53	71	79	
62	17	29	73				
73	19	71					
75	47	53	67				
78	19	41	53	67	79		
85	17	41	43	59			
88	13	17	19	37	43	59	71
90	37	43	53	71			
95	13	17	19	41	79		
99	19	43					

Figure 1-1. Primitive roots for $m = 2^{31} - 1$, $a \equiv b^c \pmod{m}$

$$(\tfrac{1}{2})^{-\sum_{i=1}^{s} d_i}$$

and $0 < l_2 < \ldots < l_s < n/q - 1$.

In practice, trinomials $x^p + x^q + 1 \pmod{2}$ are usually preferred because of the small number of shifts required to generate the sequence a; however, Tausworthe's theorem holds for any primitive polynomial $GF(2)$.

Whittlesey [25] gives an FSR RNG (feedback shift register pseudorandom number generator) algorithm due to W. B. Kendall and has shown that suitable pseudorandom numbers are obtained on the IBM 360 [$m < (n - 1)/2$]:

(1) Let register A initially contain the previous random number y in bit positions 1 to n with zero in the sign bit (position 0).

(2) Copy register A into register B and then right-shift register B m places (trinomial: $X^n \oplus X^m \oplus 1$).

(3) Exclusive-or register A into register B and also store the result back into register A.

(4) Left-shift register B $(n - m)$ positions.

Word Size	p	q	Approximate Period
16	15	4	10^4
18	17	5, 6	10^5
32	31	6, 7, 13	10^9
60	57	7, 22	10^{18}

Figure 1-2. List of p and q for primitive $x^p + x^q + 1$

(5) Exclusive-or register B into register A and zero out register A's sign bit. Register A now contains the new random number.

Payne [26] gives a FORTRAN coding of Kendall's algorithm.

The selection of shifts corresponding to primitive trinomials must be made carefully, just as one must select statistically satisfying primitive roots for HD2 from all available ones. Figure 1-2 lists only a few primitive trinomials which may be used in designing an RNG [27].

Tootill, et al. [28] warn against selecting q too small or too near $(p - 1)/2$ which results in poor run properties. In particular, n-space performance of FSR RNGs is not guaranteed unless part (3) of Tausworthe's theorem is satisfied [20].

References

[1] Hull, T. E., and Dobell, A. R. Random number generators. *SIAM Review*, 4: 230-254 (1962).

[2] Teichroew, D. Distribution sampling with high speed computers. Published Ph.D. thesis, North Carolina State College, 1953.

[3] Meyer, H. A. "Symposium on Monte Carlo Methods." New York: Wiley, 1956, p. 16.

[4] Pathria, R. K. A statistical study of randomness among the first 10,000 digits of π. *Math. Comp.*, 16 (1962).

[5] _____. A statistical analysis of the first 2,500 decimal places of e and $1/e$. *Proc. Nat. Inst. Sci. India*, a27: 270-282 (1961).

[6] Kendall, M. G., and Smith, B. B. Randomness and random sampling numbers. *J. Roy. Statis. Soc.*, 101: 162-164 (1938).

[7] Yule, G. U. A test of Tippett's random sampling numbers. *J. Roy. Statis. Soc.*, 101: 167-172 (1938).

[8] Murray, H. F. A general approach for generating natural random variables; *IEEE Trans. on Computers*, C-19(12): 1210-1213 (Dec. 1970).

[9] Lehmer, D. H. Mathematical methods in large scale computing units. *Annals Comp. Lab. Harvard Univ.*, 26: 141-146 (1951).

[10] Hull, T. E., and Dobell, A. R. Random number generators. *SIAM Review*, 4: 230-254 (1962).

[11] Murray, H. F. A general approach for generating natural random variables. *IEEE Trans. on Computers*, C-19(12): 1210-1213 (Dec. 1970).

[12] Knuth, D. E. "The art of computer programming," vol. 2: Semi-numerical algorithms. Reading, Mass.: Addison-Wesley, 1969.

[13] Westlake, W. J. A uniform random number generator based on the combination of two congruential generators. *J. ACM.*, 14: 337-310 (April 1967).

[14] Hutchinson, D. W. A new uniform pseudorandom number generator. *Comm. ACM.*, 9: 432-433 (June 1966).

[15] Kruskal, J. B. Extremely portable random number generator. *Comm. ACM.*, 12: 93-94 (Feb. 1969).

[16] Greenberger, M. An a priori determination of serial correlation in computer generated random numbers. *Math. Comp.*, 15: 383-389 (1961).

[17] Whittlesey, J. R. B. A comparison of the correlational behavior of random number generators for the IBM 360. *Comm. ACM.*, 11: 641-644 (Sept. 1968).

[18] Coveyou, R. R., and MacPherson, R. D. Fourier analysis of uniform random number generators. *J. ACM.*, 14: 100-119 (Jan. 1967).

[19] Marsaglia, G. Random numbers fall mainly on the planes. *Proc. Nat. Acad. Sci.*, 61: 25-28 (Sept. 1968).

[20] Coveyou, R. R. Random numbers fall mainly in the planes. *Review in Computing Reviews.*, 225, (April 1970).

[21] Tausworthe, R. C. Random numbers generated by linear recurrence modulo two. *Math. Comp.*, 19: 201-209 (1965).

[22] Zierler, N., and Brillhart, J. On primitive trinomials (mod 2), pt. II. *Information and Control*, 14: 566-569 (1969).

[23] _____. Primitive trinomials whose degree is a merseene exponent. *Information and Control*, 15: 67-69 (1969).

[24] Golomb, S. W. "Shift Register Sequences." San Francisco: Holden-Day, 1967.

[25] Whittlesey, J. R. B. A comparison of the correlational behavior of random number generators for the IBM 360. *Comm. ACM.*, 11: 641-644 (Sept. 1968).

[26] Payne, W. H. FORTRAN Tausworthe pseudorandom number generator. *Comm. ACM.*, 13: 57 (Jan. 1970).

[27] Whittlesey, J. R. B., and Griese, P. Multi-dimensional pseudo-random non-uniformity. Private Communication.

[28] Tootill, J. P. R., Robinson, W. D., and Adams, A. G. The runs up-and-down performance of Tausworthe pseudorandom number generators. Private Communication.

2

Pseudorandom Sequences in n-Space

2-1 Introduction

The RNGs of Chapter 1 are periodic with period m. For example, HD1 (Hull-Dobell Theorem 1) with $a = 5$, $c = -1$, $m = 8$, and $X_0 = 1$ generates the periodic sequence, $1, 4, 3, 6, 5, 0, 7, 2, 1, \ldots$ which repeats in groups of 8. These numbers may be used to randomly locate points on a directed line segment of length 8.

RNGs in n-space may be interpreted as a random selection of points from cells within a hypercube. The cellular subdivisions of such n-cubes are in general arbitrary, but for finite-memory computers, cell size (resolution) is usually limited by the word size of the host machine. Thus, in 1-space, the resolution of a full-period, m, RNG is $1/m$. Extending such a full-period RNG to the selection of points in 2-space, we see clearly that m^2 cells are required in order to maintain $1/m$ resolution. However, only m distinct numbers are supplied by a periodic RNG before repetitions occur. Thus, $m^2 - m$ cells must be empty. In general, when points are selected in n-space, m^n cells are possible, but only m are filled, leaving $m^n - m$ cells vacant.

n-CUBE THEOREM.   Let $X = \{x_i\}$ be a periodic sequence of length m and $s = \{s_i\}$ be the set of m^n grid points in an n-cube; $m \times m \times \ldots \times m$. If x is a pseudorandom sequence, m elements of s will be "randomly" selected, and $m^n - m$ elements of s will never be selected.

Figure 2-1 demonstrates this selection rule in 2-space when m equals 8.

For sufficiently large m (the length of a random sequence) and applications requiring resolution $1 \leq L \leq m$, say, it is possible to statistically expect M (number of points expected in each cell) of N (total number of points generated at random) sample points to fall in each L cell (an L-cell is a cell of size L on a side which is large enough to contain M points) if $(L/m)^n \geq M/N$ or $L/m \geq (M/N)^{1/n}$. Thus, in Figure 2-1, for instance, with $M = 5$, $N = m = 8$, $L \geq 8\sqrt{5/8}$, nearly the entire square is required. This dependence of L on n is shown in Fig. 2-2 and 2-3.

2-2 Lattices and Waves

Coveyou [8] noted the limitations imposed by the foregoing argument and theorem in a review of Marsaglia's paper [1] and theorem:

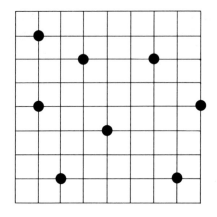

Figure 2-1. Selection of m out of m^2 cells in a 2-cube, $m = 8$

m	n			
2^{16}	73	23	16	13
2^{32}	2953	220	80	41
2^{48}	119086	2021	391	126

Figure 2-2. Values of $(n!m)^{1/n}$ of Marsaglia's theorem

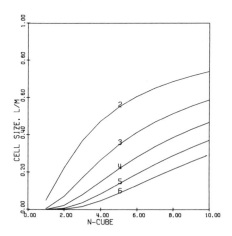

Figure 2-3. $L/m = (M/N)^{1/n}$ versus n. $N = 10^2, 10^3, \ldots, 10^6$ points are sampled with $\mu = 5$ expected in each cell.

MARSAGLIA'S THEOREM. If $c_1, c_2, \ldots, c_n$ is any choice of integers such that

$$c_1 + c_2 k + \ldots + k^{n-1} c_n \equiv 0 \quad (\text{mod } m)$$

then all the points $p_1 = (x_1, \ldots, x_{n+1})$ will lie in the set of parallel hyperplanes defined by the equations

$$c_1 z_1 + c_2 z_2 + \ldots + c_n z_n = 0, \pm 1, \pm 2, \ldots$$

There are at most $|c_1| + |c_2| + \ldots + |c_n|$ of these hyperplanes which intersect the unit n-cube, and there is always a choice of c_i, $i = 1, 2, \ldots, n_n$ such that all of the points fall in fewer than $(n!m)^{1/n}$ hyperplanes.

A table for a few values of $(n!m)^{1/n}$ is given in Figure 2-2.

Figure 2-2 shows a very low number of planes when m is small and n large. Whittlesey states [2]:

. . . these D-dimensional (n-dimensional) points will, in fact, always lie at the intersections of a number of such sets of parallel hyperplanes and hence will form space-filling lattices . . . similar to the arrangement of "atoms in a perfect crystal at absolute zero" . . . the granularity . . . may be thought of as imposing an upper bound on the "effective" number of meaningful bits available when performing calculations requiring multi-dimensional uniformity.

The series of Figures 2-4 and 2-5 show this growth of "crystals" as more and more points are taken from the RNG, $X_{i+1} \equiv 17 X_i - 1 \pmod{512}$. The low period, 512, enables one to easily see the lattice appear; however, if an RNG of very long period, say 2^{31}, were employed, the lattice structure would be too fine to see, as in the series of Figures 2-6 and 2-7.

Unfortunately, the period, m, of congruential RNGs is dependent upon fixed computer word size. A possible remedy is to operate a composite bank of RNGs, one for each dimension in n-space. Figures 2-8 and 2-9 were plotted from two RNGs with multipliers 17 and 33. In each RNG, the starting values were $X_0 = 1$ and $c = -1$. Clearly, the lattice structure is gone, but a wavelike clustering of points persists. The waves are caused by synchronized runs-up-runs-down patterns in the two generators.

Coveyou and MacPherson [3] performed Fourier analysis on single RNGs which is extended here to investigate a bank of Lehmer RNGs.

COVEYOU-MACPHERSON (CM) THEOREM. The finite Fourier transformation of the pattern function

$$f(t_1, \ldots, t_n) = \lim_{m \to \infty} \frac{1}{m} \sum_{0 \le k \le m} \delta_{x_k t_1} \delta_{x_{k+1} t_2} \cdots \delta_{x_{k+n-1} t_n}$$

$$X_k \equiv a X_{k-1} + c \quad (\text{mod } m) \qquad \delta_{rs} = \begin{cases} 1 & r = s \\ 0 & r \neq s \end{cases}$$

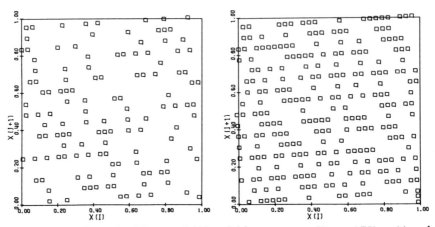

Figure 2-4. Growth of "crystals" in a 9-bit generator: $X_{i+1} \equiv 17X_1 - 1 \pmod{512}$. (*a*) 128 points, (*b*) 256 points.

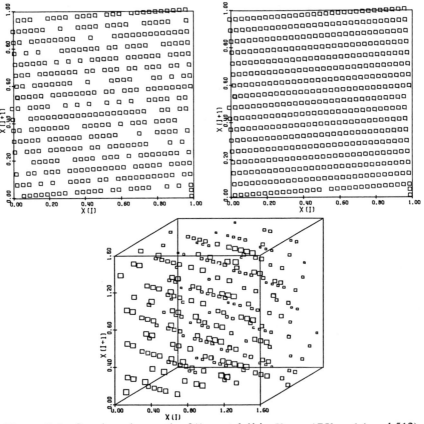

Figure 2-5. Continued growth of "crystals" in $X_{i+1} \equiv 17X_i - 1 \pmod{512}$; (*a*) 384 points; (*b*) 512 points; (*c*) plot in 3-space of 256 successive triplets.

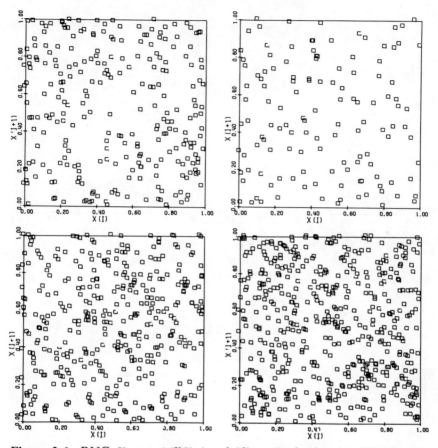

Figure 2-6. RNG $X_{i+1} \equiv 14^{29}X_i \pmod{2^{31} - 1}$ plotting (*a*) 128, (*b*) 256, (*c*) 384, and (*d*) 512 successive pairs.

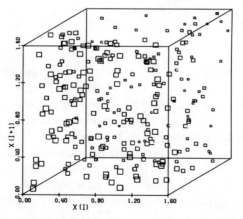

Figure 2-7. 3-space plot of RNG $X_{i+1} \equiv 14^{29}X_i \pmod{2^{31} - 1}$ and 256 successive 3-tuples.

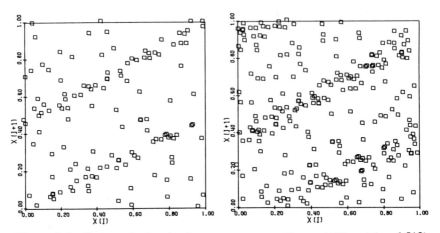

Figure 2-8. Composite bank of two generators, $X_{i+1} \equiv 17X_i - 1 \,(\mathrm{mod}\ 512)$, $Y_{i+1} \equiv 33Y_i - 1 \,(\mathrm{mod}\ 512)$. Each pair $(X_{i+1},\ Y_{i+1})$ is plotted (a) 128 and (b) 256 times as $(X_i,\ X_{i+1})$.

is given by

$$F(s_1, \ldots, s_n) = \sum_{0 \le t_1 \le \ldots \le t_n \le m-1} \exp\left[\frac{-2\pi i}{m} \left(\sum_{j=1}^{n} s_j t_j \right) \right]$$
$$\cdot f(t_1, \ldots, t_n)$$

and yields a minimum wave number

$$\nu_n = \min_{s_i} \sum_{i=1}^{n} s_i^2$$

subject to the constraint

$$\sum_{i=1}^{n} a^{i-1} s_i \equiv 0 \quad (\mathrm{mod}\ m),\ s_j \ne 0.$$

Furthermore, $\nu_n \le \gamma_n m^{1/n}$ where γ_n are lattice constants [4]

The pattern function f merely counts the number of times a particular n-tuple appears in the sequence $\{X_k\}$. Waves represented by wave numbers ν_n are interpreted as cyclic contractions/rarefactions in the density of points in n-space. An analogy with Marsaglia's lattices may be valid if we think of Marsaglia's planes as lying on the crest/trough of waves [5]. Figure 2-10 is a comparison in terms of bits of accuracy of Marsaglia's formula and the bound of CM (Coveyou-MacPherson theorem). Note, the CM bound is much stricter as n increases. In practice, the CM bound is more indicative of actual behavior.

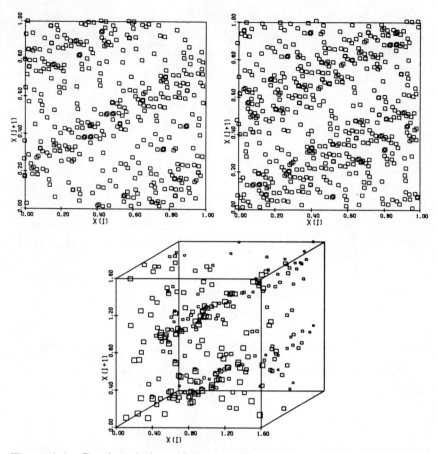

Figure 2-9. Continued plots of $X_{i+1} \equiv 17X_i - 1$, $Y_{i+1} \equiv 33Y_i - 1 \pmod{512}$ for (a) 384 and (b) 512 pairs. The 3-tuple X_{i+1}, Y_{i+1}, X_{i+2} is plotted for 256 points in the cube of (c).

It may be desirable to disrupt the lattice regularity or wave structure of congruential RNGs shown to exist by Marsaglia and by Coveyou and MacPherson. An immediate modification which uses a composite bank of n RNGs for n-space will serve this purpose, if the following theorem is considered.

MULTI-RNG THEOREM.   A bank of n congruential generators

$$
\begin{aligned}
X_{i+1} &\equiv a_x X_i + c_x &&\pmod{m} \\
Y_{i+1} &\equiv a_y Y_i + c_y &&\pmod{m} \\
&\;\;\vdots \\
Z_{i+1} &\equiv a_z Z_i + c_z &&\pmod{m}
\end{aligned}
$$

Dimension n	Lattice Constants	CM Wave Number	Marsaglia Planes
1	1.000	31.00	31.00
2	1.155	15.60	16.00
3	1.260	10.44	11.19
4	1.414	7.88	8.90
5	1.516	6.32	7.58
6	1.665	5.29	6.75
7	1.811	4.55	6.19
8	2.000	4.00	5.79
9	2.000	3.56	5.50
10	1.792	3.18	5.28
16	—	—	4.70

Figure 2.10. Table of Marsaglia and Coveyou-MacPherson bits of accuracy for $m = 2^{31}$.

will produce a sequence with period nm but with

$$\lim_{a_x, a_y, \ldots, a_z \to \sqrt{m}} \nu_n = 0$$

Proof. The derivation is given for $n = 2$ without loss of generality.

Since X_k and Y_k are periodic with period m and since $f(t_1, t_2) = 0$ unless

$$(t_1, t_2) = (X_k, Y_k),$$

$$F(s_1, s_2) = \sum_{k=0}^{m} A \exp [(-2\pi i/m)S(a)].$$

$A = \exp [(-2\pi i/m) (s_1 C_x + s_2 C_y)]$ and $S(a) = a_x s_1 + a_y s_2$. Summing in closed form, note that $|F(s_1, s_2)| = 0$ unless $S(a) \equiv 0 \pmod m$. The smallest wave number, $\nu_2 = \min_{s_1, s_2} \sqrt{s_1^2 + s_2^2}$ is estimated by a continuous analog as follows:

$$\psi = s_1^2 + s_2^2 - \lambda(a_x s_1 + a_y s_2 - km)$$

setting

$$\frac{\partial \psi}{\partial s_1} = \frac{\partial \psi}{\partial s_2} = \frac{\partial \psi}{\partial \lambda} = 0$$

and solving,

$$s_1 = \frac{a_x}{a_x^2 + a_y^2} km$$

$$\lambda = \frac{2km}{a_x^2 + a_y^2}$$

$$s_2 = \frac{a_y}{a_x{}^2 + a_y{}^2} \, km$$

Hence, $v_2 = k'm^2/(a_x{}^2 + a_y{}^2) \pmod{m}$ for some k'. Now, if $a_x = a_y = \sqrt{m}$, then $v_2 \equiv 0 \pmod{m}$, and since a_x, $a_y \leq m - 1$, the theorem follows.

Figure 2-11 demonstrates "repaired" sequences obtained with multipliers far from $\sqrt{m}$, as predicted. Thus, the advice of Greenberger [6] that multipliers near $\sqrt{m}$ yield favorable correlation properties is definitely to be avoided when using a bank of HD1 or HD2 RNGs. Further, note that two generators with the same starting values and same multipliers will give very correlated sequences, but if the multipliers differ slightly, good results will be realized as long as multipliers are far from $\sqrt{m}$.

2-3 Feedback Shift Register Sequences

FSR RNGs of Tausworthe's theorem represent a fresh approach to RNG theory based on shifts and exclusive-or's. Figure 2-12 demonstrates the growth of a uniform covering of 2-space. The striking breakdown shown in Figure 2-12d is explained by Tootill, et al [7]. For shifts close to $(p - 1)/2$ or very near 1, the Kendall alogorithm produces sequences with long runs-up or runs-down. The situation in 3-space of Figure 2-12d is quite similar to the matched-runs patterns of the multi-Lehmer RNGs. Runs artificially induce a linear correlation between X_{i+1} in the two types of RNGs. Excessive run lengths show up as groupings in a $2, 3$-space plot. A closer examination of runs-up of HD1 and HD2 and FSR sequences will be made later.

Restricting shift lengths, $1 < q < p-1$, and fixing p as large as possible for a given computer, a period of $m = 2^p - 1$ is obtainable without serious runs-up-runs-down deficiency. Recall, however, that $m^n - m$ vacancies will occur in n-space for such periodic sequences. The question naturally arises, what limitation is imposed on n-space RNGs with period m? The key to this question lies in the fact that the period m is based on *nonrepeatability* of any of the m numbers. This restriction is too great for large n. Fortunately, there are $m!$ permutations of m numbers, leading one to seek extended sequences with period $m(m!)$ (allowing repeats). If the full $m(m!)$ extended period is exploited, then n is restricted by

$$m^n \leq m(m!)$$

Using an approximation for large m,

$$\log m! \approx \int_1^m \log x \, dx = m \, (\log m) - m$$

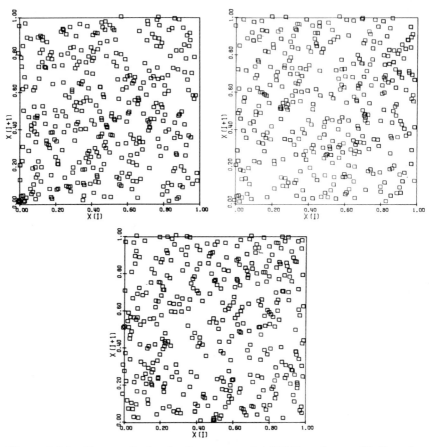

Figure 2-11. Composite bank of congruential RNGs using multipliers far from $\sqrt{m}$: (a) 5, 9; (b) 257, 509; (c) 257, 9, and 384 points.

which has error $\delta (\log m)$, by Stirling's formula,

$$n \le m \left(1 - \frac{1}{\log m} \right) + 1$$

Thus, very large values of n are possible while all m^n cells of an n-cube are filled.

EXTENDED SEQUENCE THEOREM   A basic sequence, $x = \{x_i\}$ of period m can be extended to a sequence of length $m(m!)$ if each x_i is allowed to appear $m!$ times. Furthermore, for n-space, require

$$n < m \left(1 - \frac{1}{\log m} \right) + 1$$

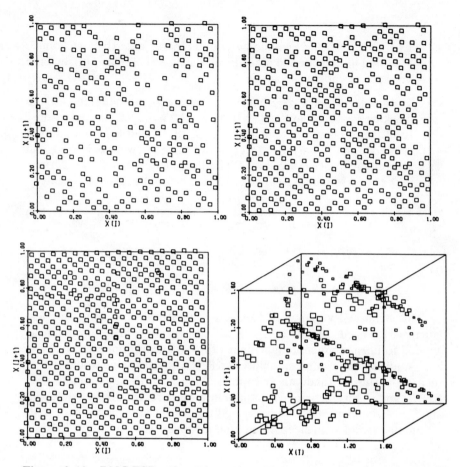

Figure 2-12. RNG FSR: $X^9 + X^4 + 1$. (*a*) 127, (*b*) 255, (*c*) 511 points plotted successive pairs; (*d*) 255 successive 3-tuples plotted in 3-space.

Extended sequences offer hope of curing the evils of the *n*-cube theorem only if ways of efficiently generating extended sequences are found. The multi-Lehmer RNG is only an *n*-fold extended sequence when m^n-fold increase is needed. A masked FSR sequence extends the period of a basic FSR sequence without altering the basic RNG properties [8].

MASKED FSR (MFSR) SEQUENCE THEOREM. The sequence $\{a_i\}$ generated by the FSR, $X^p + X^q + 1$ of maximal period $m = 2^p - 1$ is extended to the sequence $\{b_i\}$ generated by

$$b_i = M_k \oplus a_i \quad i, k = 0, 1, \ldots, m - 1; i \neq k$$

$$b_i = a_i \quad i = k$$

where $M_k = a_k$, and k is a mod m counter incremented by 1 each time i $\equiv 0 \pmod{m}$. The period of $\{b_i\}$ is m^2, the mean is $E(b) \equiv E(a)$, and the variance is var $(b) \equiv$ var (a).

Proof.   Initially set $M_0 = a_0$ (p bits) and generate $\{b_i\}$ from $\{a_i\}$. If $i = k$, the masking would yield zero (p bits), so M_i is never masked against itself. Continuing, increment mask M_0 to M_1 and generate another sequence of $\{a_i\}$, only in different order since a new mask is used, and all m elements of $\{a_i\}$ are unique. Finally, m sequences, all reorderings of the basic sequence, will be generated before the cycle is repeated.

$$E(b) \quad = \quad \frac{1}{m^2} \sum_{k=0}^{m-1} \sum_{i=0}^{m-1} a_i = \frac{1}{m} \sum_{i=0}^{m-1} a_i = E(a) \qquad (2.1)$$

$$E(b^2) \quad = \quad \frac{1}{m^2} \sum_{k=0}^{m-1} \sum_{i=0}^{m-1} a_i^2 = \frac{1}{m} \sum_{i=0}^{m-1} a_i^2 = E(a^2) \qquad (2.2)$$

$$\text{var } (b) \quad = \quad E(b^2) - E^2(b) = E(a^2) - E^2(a) = \text{var } (a) \qquad (2.3)$$

The mean and variance of MFSR are unaltered, and only the cycle length is changed. MFSR is a special case of affine transformation as applied to linear sequential machines [9]. Statistical properties have not been studied to determine suitability as a RNG; however, one would expect their properties of randomness to differ little from those of simple FSR RNGs, and so this technique is of little interest here. In addition, note that all $2^p - 1$ numbers must be generated before repeats can be expected. Fractional period sequences, then, hold no advantage over other RNGs.

2-4 Conclusions

The theoretical results just presented indicate the need for RNG with very great period and/or increased word size. Since word size is fixed for most computers, an increase in period is sought. Also, repeatability of numbers within a full period is required to maintain n-space uniformity by filling m^n cells. Empirical tests must be applied to determine damage, if any, due to lattice regularity.

Thus, a new generator with repeatability within a full period is presented. This RNG has the extended sequence property for n-space uniformity.

References

[1] Marsaglia, G. Random numbers fall mainly on the planes. *Proc. Nat. Acad. Sci.*, 61: 25-28 (Sept. 1968).

[2] Whittlesey, J.R.B., and Griese, P. Multi-dimensional pseudo-random non-uniformity. Private Communication.

[3] Coveyou, R.R., and MacPherson, R.D. Fourier analysis of uniform random number generators. *J. ACM.*, 14: 100-119 (Jan. 1967).

[4] Cassels, J.W.S. "An Introduction to the Geometry of Numbers." Berlin: Springer-Verlag, 1959, p. 332.

[5] Whittlesey, J.R.B. On the multidimensional uniformity of pseudo-random generators. *Comm. ACM.*, 12: 247 (May 1969).

[6] Greenberger, M. An a priori determination of serial correlation in computer generated random numbers. *Math. Comp.*, 15: 383-389 (1961).

[7] Tootill, J.P.R., Robinson, W.D., and Adams, A.G. The runs up-and-down performance of Tausworthe pseudo-random number generators. *J.ACM.*, (1971). Private Communication.

[8] Coveyou, R.R. Random numbers fall mainly in the planes. *Review in Computing Reviews* 225 (April 1970).

[9] Wang, K.C. Transition graphs of affine transformations on vector spaces over finite fields. *J. Franklin Institute*, 283: 55-72 (Jan. 1967).

3

Generalized Algorithm for FSR

3-1 Introduction

Kendall's algorithm (Chapter 1) selects successive n-tuples from the basic sequence $\{a_i\}$, where $a_k = a_{k-p+q} + a_{k-p}$, $k = p, p + 1, \ldots$, given $a_{p-1}, \ldots$, a_0, and FSR, $x^p + x^q + 1$. For example, $x^4 + x^1 + 1$ and $a_0 = a_1 = a_2 = a_3 = 1$ yields $\{a_i\}_0^{14} = 111100010011010$, and selecting 4-tuples by Kendall's algorithm produces the random numbers in Figure 3-1 ($n = 4$, initial register $A = 1111$).

Note that each column of bits in Figure 3-1 is a delayed replica of the basic sequence $\{a_i\}$. And, since each column obeys the basic sequence recurrence relation, the 4-bit words themselves also obey the recurrence relation. The sequence of 4-tuples is generated by applying the recurrence relation to 4-tuples. That is, $w_k = w_{k-p+q} \oplus w_{k-p}$, $k = p, p + 1, \ldots$, given p starting words. Proceed with caution, however, since delayed replicas will not always be produced by Kendall's algorithm. In fact, delayed replicas are guaranteed only for word sizes, $L = 2^k$, $k = 1, 2, \ldots$. To see this, repeatedly square the primitive polynomial $p(x) = x^p + x^q + 1$, mod 2. The resulting polynomial after squaring k times is $x^{2^k p} + x^{2^k q} + 1$, or $y = x^{2^k}$. $p(y) = y^p + y^q + 1$, mod 2. The sequence produced by $p(y)$ is identical to that produced by $p(x)$, but $p(y)$ is obtained by skipping every (2^k)th bit-column sequence instead of selecting bits consecutively (row sequence). Hence, the sequence obtained by going down a column rather than row by row is equivalent to a sequence obtained by skipping every Lth bit in the basic sequence $a = \{a_i\}$. In Figure 3-1, $L = 4 = 2^2$, and so the high-order bit column is obtained by skipping every fourth bit in $\{a_i\}_0^{14}$, starting with bit $a_3 = 1$.

If Kendall's algorithm is abandoned, but the idea of using delayed replicas to construct, column by column, a sequence of random numbers is not, then a new algorithm is possible. Therefore, the relative delay between columns of bits may be arbitrary, giving rise to many sequences of words, $\{w_i\}$, in addition to the single sequence of Kendall's algorithm.

Actually, there are $Q(m)$ distinct basic sequences which may be used [1]:

$$Q(m) = \phi(m)/P$$

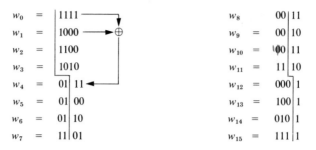

w_0 =	1111	
w_1 =	1000	
w_2 =	1100	
w_3 =	1010	
w_4 =	01 11	
w_5 =	01 00	
w_6 =	01 10	
w_7 =	11 01	

w_8	00 11
w_9 =	00 10
w_{10} =	00 11
w_{11} =	11 10
w_{12} =	000 1
w_{13} =	100 1
w_{14} =	010 1
w_{15} =	111 1

Figure 3-1. Kendall's algorithm for the polynomial $x^4 + x^1 + 1$

$\phi(m)$ = Euler function; the number of integers $< m$ relatively prime to m

$m = 2^p - 1$ is the period of FSR

p = degree of primitive polynomial

In particular, when m is prime, $Q(m) = (m - 1)/p$. Also a reversed basic sequence is obtainable from the reciprocal polynomial $(x^p + x^{p-q} + 1)$.

The period of the sequence $\{w_i\}$ is determined by the polynomial, while the order of the sequence is determined by the p starting words. Starting values may be given by Kendall's algorithm or by choosing any relative delay between columns. With both approaches, a full period, $m = 2^p - 1$ sequence, $\{w_i\}$, is guaranteed when the polynomial $x^p + x^q + 1$ is primitive, and statistically independent columns are used in starting the sequence. When each column is a delayed replica of $\{a_i\}$, statistical independence of p-tuples is assured.

The GFSR (generalized feedback shift register) algorithm is as follows:

GFSR: 0. If $k \neq 0$, go to 2 (k initially zero).

1. Initialize, $w_0, \ldots, w_{p-1}$ using a delayed basic sequence, $\{a_i\}$, to obtain each column of $w_0, \ldots, w_{p-1}$.

2. $k \leftarrow k + 1$

3. If $k > p$, set $k \leftarrow 1$.

4. $j \leftarrow j + q$.

5. If $j > p$, set $j \leftarrow j - p$.

6. Exclusive-or $w_k + w_j$.

7. Store $w_k \leftarrow w_k + w_j$.

A rotating table of p words is kept in GFSR, which is implemented as a FORTRAN function in Figure 3-2.

```
0001            FUNCTION RAND(M,L,SHIFT,WDSIZE)
         C
         C      M(L)=TABLE OF L PREVIOUS RANDOM NUMBERS.
         C      L,SHIFT=POLYNOMIAL PARAMETERS: X**L+X**SHIFT+1.
         C      NOTE: TWO'S COMPLEMENT ARITHMETIC IS DONE IN ASCII FORTRAN...
         C      IN PLACE OF LOGICAL .NOT. OPERATOR.
         C      WDSIZE=WORD SIZE (BITS) OF HOST MACHINE.
         C
0002            LOGICAL AA,BB,LCOMPJ,LCOMPK
0003            INTEGER A,B,SHIFT,WDSIZE,M(1)
0004            EQUIVALENCE (AA,A),(BB,B),(MCOMPJ,LCOMPJ),(MCOMPK,LCOMPK)
0005            DATA J/0/
0006            N=(2**(WDSIZE-1)-1)*2+1
0007            J=J+1
0008            IF(J.GT.L) J=1
0009            K=J+SHIFT
0010            IF(K.GT.L) K=K-L
0011            MCOMPJ=N-M(J)
0012            MCOMPK=N-M(K)
0013            A=M(K)
0014            B=M(J)
0015            BB=LCOMPJ.AND.AA.OR.LCOMPK.AND.BB
0016            M(J)=B
0017            RAND=FLOAT(M(J))/FLOAT(N)
0018            RETURN
0019            END
```

Figure 3-2. FORTRAN implementation of GFSR algorithm. Initialization is assumed done by SETR in Figure 3-3.

3-2 Initialization of GFSR Algorithm

The GFSR algorithm is self-initializing in the sense that delayed replicas are produced by the same procedure that generates full words. Statistical independence of starting columns is guaranteed by delayed replicas if delays are less than full period. (If p is relatively prime to $2^p - 1$, then delays exceeding the period will produce a delayed replica also.) That is, every p-tuple is generated (except all zeros) before any p-tuple repeats. Each column is a p-tuple, and therefore must be a different pattern than all others. For example, in Figure 3-1, initialization can be done from most significant to least significant bits (left to right) starting with {1111}. The recurrence relation is applied 11 times in the example to get the next column {1010}. A second 11 applications results in {1001}, and finally {1000} is obtained.

The recurrence is applied by calling the GFSR RNG with zero fills placed to the right. Each new column is shifted to the right after being generated, and the original column (1111) is replaced at the extreme left column. In more realistic (bigger word size) generators, the full period will not be exhausted by initialization, in keeping with conditions for statistical independence (here, independence is implied because 11 is relatively prime to 15).

For ease in implementation, the (111...) starting p-tuple is used in SETR as given in Figure 3-3. However, SETR applies $5000p$ additional delays, so that the leading (111...) column is "recurrenced" also, thus giving a random pattern of leading bits rather than all 1s. This additional "recurrencing" is necessary to allow (111...) to "die out." The effects of such a regular pattern carry over to later p-tuples. For example, (111...), $p = 98$, is transformed to 72 zeros, 27 ones; next to 45 zeros, 53 ones; etc. The grouping of all 1s or all 0s becomes increasingly smaller and more "random" through repeated application of the recurrence relation. "Damping" of the initial p-tuple is done in SETR by applying the recurrence relation $5000p$ times to full word starting values. It should be noted that if generality relative to word size is not desired, then additional "recurrencing" is not necessary when initialization is performed from least significant to most significant bits. The most significant bit will have been delayed $L \times$ delay times since it is "recurrenced" once for each bit in an L-bit word. Thus, the additional "randomizing" of the initial (111...) pattern will have been done without additional labor.

Finally, SETR returns a value for the number of statistically independent columns available to the initialization procedure. The statistical quality of these delays must be tested, however, as shown later in Figure 3-6.

3-3 Generality of GFSR

The parallel nature of GFSR immediately generalizes to L-bit machines, independent of the relation between L and p. Thus, for $L < p$, many repeated numbers will occur, but cycle length m is still $2^p - 1$. Figure 3-4 demonstrates the case $L = 2$, and $x^4 + x^1 + 1$ from Figure 3-1. Here, 2^{p-L} nonzero duplicates and $2^{p-L} - 1$ zeros are produced in one full period.

Very long period sequences can be generated on any L-bit machine merely by selecting p large. A partial table of primitive polynomials of large p is reproduced in Figure 3-5. A complete table can be found in reference [2].

3-4 Period of GFSR

An "unlimited" period is possible without increasing the word size of host machines. For example, $m = 2^{532} - 1$ is obtained by using $x^{532} + x^{37} + 1$ from Figure 3-5. To exhaust this cycle would require many years on a very fast computer; i.e., if 10^6 numbers per second were generated, approximately 10^{150} years would be needed to complete the cycle! More important, though, is the repeatability of numbers within a full period. Thus, an extended sequence is obtained with desirable n-cube theorem properties (see Chapter 2).

```
0001          INTEGER FUNCTION SETR(M,L,DELAY,SHIFT,WDSIZE)
       C
       C      SETR=COLUMN NUMBER OF REPEATING ONE PATTERN.  IF SETR< OR = L,
       C      THEN AN IMPROPER SHIFT LENGTH HAS BEEN SELECTED.
       C      M(L)=TABLE OF L PREVIOUS RANDOM NUMBERS.
       C      L,SHIFT=POLYNOMIAL PARAMETERS: X**L+X**SHIFT+1.
       C      DELAY=RELATIVE DELAY BETWEEN COLUMNS OF M(L) , IN BITS.
       C      WDSIZE=WORD SIZE (BITS) OF HOST MACHINE.
       C
0002          INTEGER DELAY,SHIFT,ONE,WDSIZE,M(1)
0003          SETR=L+1
0004          ONE=2**(WDSIZE-1)
0005          DO 1 I=1,L
0006     1    M(I)=ONE
0007          DO 4 K=1,WDSIZE
0008          DO 2 J=1,DELAY
0009     2    X=RAND(M,L,SHIFT,WDSIZE)
0010          KOUNT=0
0011          DO 3 I=1,L
0012          ITEMP=ONE/2**(K-1)
0013          ITEMP=(M(I)-M(I)/ONE*ONE)/ITEMP
0014          IF(ITEMP.EQ.1) KOUNT=KOUNT+1
0015          IF(K.EQ.WDSIZE) GO TO 3
0016          M(I)=M(I)/2+ONE
0017     3    CONTINUE
0018          IF(KOUNT.EQ.L) SETR=K
0019     4    CONTINUE
0020          DO 5 I=1,5000
0021          DO 5 J=1,L
0022     5    X=RAND(M,L,SHIFT,WDSIZE)
0023          RETURN
0024          END
```

Figure 3-3. FORTRAN implementation of SETR to initialize GFSR algorithm.

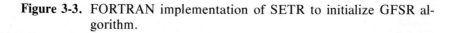

Starting values

$w_0 = 11$ $w_8 = 00$
$w_1 = 10$ $w_9 = 00$
$w_2 = 11$ $w_{10} = 10$
$w_3 = 10$ $w_{11} = 11$
$w_4 = 01$ $w_{12} = 00$
$w_5 = 01$ $w_{13} = 10$
$w_6 = 01$ $w_{14} = 01$
$w_7 = 11$ w_{15} repeats

Figure 3-4. GFSR algorithm for $L = 2$, polynomial $x^4 + x^1 + 1$

3-5 Mean, Variance, and Correlation of GFSR

The theoretical *mean* and variance of a GFSR sequence are guaranteed by periodicity:

$$u = \frac{1}{m} \sum_{i=0}^{m-1} w_i \quad m = 2^p - 1$$

p	q
47	5, 14, 20, 21
95	11, 17
98	11, 27
111	10, 49
124	37
170	23
250	103
380	47
476	15, 141
532	37

Figure 3-5. Primitive polynomials, $x^p + x^q + 1$, p large.

For an L-bit machine, 2^{p-L} nonzero duplicates and $2^{p-L} - 1$ zeros will be generated before the entire sequence repeats:

$$u = \frac{2^{p-L}}{m} \sum_{i=1}^{2^L-1} i + \frac{2^{p-L} - 1}{m} \sum_{i=1}^{2^L-1} 0$$

$$= \frac{2^{p-L}}{m} \left(\frac{2^{p-L}}{2} - 1 \right) \approx \frac{1}{2} \frac{2^{p-L}}{m}$$

In the normalized $(0, 1)$ interval, $u_0 \approx 1/2$.

The *variance* is

$$\sigma^2 = \frac{1}{m} \sum_{i=0}^{m-1} w_i^2 - u^2 = \frac{(2^{p-L})^2}{m} \sum_{i=1}^{2^L-1} i^2 - u^2$$

$$\approx \frac{1}{3} \frac{2^{p+2L}}{m} - u^2$$

and normalized to $(0, 1)$, $\sigma_0^2 \approx 1/3 - 1/4 = 1/12$.

The *correlation* obtained by averaging over the entire period

$$E[R(t)] = \frac{1}{m} \sum_{j=0}^{m-1} \frac{1}{N} \sum_{i=0}^{N-1} w_i w_{i+t}$$

can be derived by using techniques identical to Tausworthe's [3]. However, intuition indicates that columns which have nearly the same delay and hence are nearly equal, bit by bit, should result in large correlation coefficients. Intuitively, the correlation must decrease as relative delay increases. Full period analysis by Tausworthe does not properly model this microstructure and tells us nothing about short sequences. Therefore, an empirical rule is used which computes the maximum correlation coefficient over a range $0 \le t \le 50$. A plot of maximum correlation coefficient versus relative delay between columns shown in Figure 3-6 indicates a relative column delay of $100p$ or more to be satisfactory. It is possible to find smaller delays which also give satisfactory correlation, but several polynomials were tested and all were found to be "safe" with delays of order $100p$. Finally, selection of delays which are multiples of p assures linear independence of starting values discussed in the initialization of GFSR algorithm.

3-6 Multidimensional Uniformity of GFSR

Figure 3-7 shows a much improved 9-bit generator when compared with Lehmer and Kendall RNGs shown in Chapter 2. The underlying reason is, of course, the longer than $m = 2^9$ period, as well as the repeatability of numbers within one cycle of the generator.

According to the n-cube theorem stated in Chapter 2, to fill in m^n cells in n-space, $(2^L)^n < 2^p - 1$, or $nL \le p$. Therefore, a necessary condition for n-space uniformity is that $n \le p/L$. For example, suppose $L = 15$ and $p = 98$; then uniformity may be possible up to dimension $n = 6$.

n-space uniformity cannot be guaranteed without knowledge of the order of numbers generated by GFSR. Once this order is known, the sequence is no longer random—it is predictable. This paradox forces us to rely on empirical tests to determine uniformity to a greater extent.

GFSR THEOREM.   The sequence of L-bit numbers generated by GFSR, $x^p + x^q + 1$, $p \ge L$, has

(1) Period $m = 2^p - 1$, if $x^p + x^q + 1$ is primitive.

(2) Normalized mean $u_0 \approx 1/2$.

(3) Normalized variance $\sigma_0^2 \approx 1/12$.

(4) Potential n-space uniformity for $n \le p/L$.

3-7 Conclusions

The advantages of GFSR RNG are summarized:

(1) *Speed*: one exclusive-or versus a multiply/reduction modulo m for Lehmer, and two exclusive-ors/two shifts for Kendall's algorithm.

(2) *Generality*: a standard FORTRAN subprogram can be implemented on any computer, regardless of its word size. A small word size merely

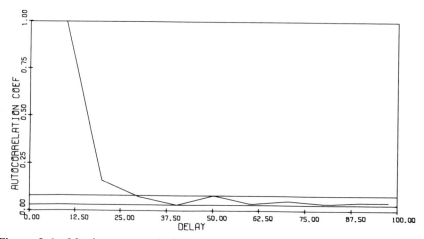

Figure 3-6. Maximum correlation versus relative column delay (measured in p units) for GFSR RNG using 15 bits, $X^{98} + X^{27} + 1$

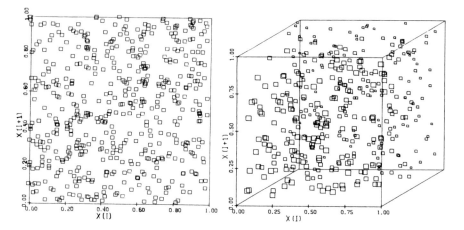

Figure 3-7. (*a*) Two-dimensional plot of GFSR, $X^{31} + X^{13} + 1$, 9-bit word size and delay of 93; (*b*) three-dimensional plot of GFSR, $X^{31} + X^{13} + 1$, delay=93.

reduces the resolution of random numbers produced, but high-order bits will be unchanged on any machine. Comparisons with sequences obtained on other machines using a FORTRAN program are given in Appendix A for generators.

(3) *Unlimited period*: any primitive polynomial can be implemented when sufficient memory is available for storage of p words. For example, x^{98}

MODEL 4 MICRO CODE

```
                    ORG  X'428'
0428  5004   RND    L    RAH,H(RND)
0429  3080          C    SB
042A  4453          L    MR4,MAR          SAVE TABL IN MR4.
042B  43A3          L    MR3,MDR          SAVE I IN MR3.
042C  5802          L    AR,X'2'
042D  C550          A    MAR,MAR,NF+NC
042E  3100          C    MR               GET S,N: LOC TABL+2.
042F  C050          A    MR0,MAR,NF+NC    SAVE TABL+4 IN MR0.
0430  58FF          L    AR,X'FF'
0431  82A3          N    MR2,MDR,NF       MASK-CFF S; N IN MR2.
0432  48AF          L    AR,MDR,C S
0433  91FF          N    MR1,X'FF'        CROSS SHIFT AND MASK-C
0434  4833          L    AR,MR3           COMPARE I:N.
0435  E824          S    AR,MR2,NC
0436  1138          B    L,OK             BRANCH IF I<N.
0437  5300          L    MR3,X'00'        OTHERWISE, SET I=0.
0438  4813   OK     L    AR,MR1
0439  C130          A    MR1,MR3,NF+NC    J=I+S
043A  4813          L    AR,MR1
043B  E224          S    MR2,MR2,NC       (MR2)=J-N.
043C  113E          B    L,OK2            BRANCH IF J<N.
043D  4123          L    MR1,MR2          OTHERWISE, SET J=J-N.
043E  4818   OK2    L    AR,MR1,SL+NC     D1=2*J.
043F  C500          A    MAR,MR0,NF+NC    LOC TABL+D1+4.
0440  3100          C    MR               GET TABL(J).
0441  4838          L    AR,MR3,SL+NC     D2=2*I.
0442  C500          A    MAR,MR0,NF+NC    LOC TABL+D2+4.
0443  48A3          L    AR,MDR           SAVE TABL(J).
0444  3100          C    MR               GET TABL(I).
0445  AEA3          X    YD,MDR,NF        EX-OR TABL(J)+TABL(I).
0446  4AE3          L    MDR,YD
0447  3200          C    MW               RESULT IN TABL(I).
0448  5801          L    AR,X'01'         INCREMENT.
0449  CA30          A    MDR,MR3,NF+NC    I=I+1.
044A  4543          L    MAR,MR4
044B  3200          C    MW               NEW I IN LOC TABL.
044C  4563          L    MAR,LOC
044D  066B          D    LOC,LOC,P2N
                    END
```

Figure 3-8. Interdata 4 microprogram for GFSR. Execute, RNG REG1, TABLE. The resulting random integer is returned to register REG1, and the address of TABLE locates the memory table.

$+x^{27} + 1$ can be used on a 15-, 24-, 31-, 35-, or 47-bit machine and a cycle length of $2^{98} - 1$ can be realized on them all.

Moreover, the speed/cost ratio is further enhanced when GFSR is implemented as a microprogrammed instruction. Figure 3-8 presents a microprogram for computing a pseudorandom number each time RNG REG1, TABLE is issued. The Interdata 4 machine instruction, $D4_{16}$ is wired into read-only memory, as shown in Figure 3-9.

Impressive speed (24 microseconds on the Interdata 4, which is equivalent to 5 to 6 microseconds on an IBM 360/65) and extended period are realized on a small-word machine (15 bits). The period of $2^{98} - 1$ greatly surpasses the previously attainable period of $2^{15} - 1$.

34

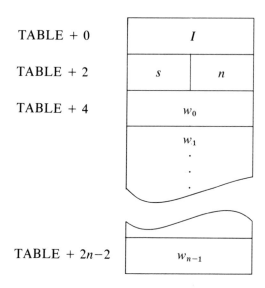

TABLE + 0	I	
TABLE + 2	s	n
TABLE + 4	w_0	
	w_1	
	$\vdots$	
TABLE + $2n-2$	w_{n-1}	

Figure 3-9. Organization of TABLE for microprogram RND. GFSR uses $x^n + x^s + 1$ by exclusive-or of w_n with w_{n+s}.

The table for RNG REG1, TABLE is organized as shown in Figure 3-9. Thus $x^n + x^s + 1$ is generated using index I.

References

[1] Whittlesey, J.R.B., and Griese, P. Multi-dimensional pseudo-random non-uniformity. Private Communication.

[2] Zierler, N., and Brillhart, J. On primitive trinomials (mod 2), pt. II. *Information and Control*, 14: 566-569 (1969).

[3] Tausworthe, R.C. Random numbers generated by linear recurrence modulo two. *Math. Comp.*, 19: 201-209 (1965).

4   **Testing RNGs**

4-1 Introduction

The deterministic nature of all RNGs makes it possible to devise particular tests which condemn a particular RNG. According to Hull and Dobell [1]:

. . . no finite class of tests can guarantee the general suitability of a finite sequence of numbers. Given a set of tests, there will always exist a sequence of numbers which passes these tests but which is completely unacceptable for some particular application.

The goal in designing an RNG is to simulate as closely as possible the statistical behavior of a uniform variate.

Analytic formulas computed over the entire cycle of periodic sequences are necessary, but not sufficient, conditions for uniformity, since a full period sequence is rarely used in practice. Therefore, analytic results are desirable when possible, but empirical tests more convincing.

In the following, both analytic and empirical tests will be used to investigate behavior of RNGs in low-dimensional and high-dimensional spaces. The suitability, then, of a particular RNG for a particular application will be determined on the basis of tests passed.

The classical tests of Kendall and Smith [2, 3], recent tests of MacLaren and Marsaglia [4], and others [5, 6] are applied to Lehmer, Tausworthe, and general FSR pseudorandom number generators.

4-2 Low-dimensional Tests

Perhaps the most important RNG parameters to be considered are mean, M, and Variance, σ^2. For any full period m, RNG, each integer, $i = 1, 2, \ldots,$ m must appear once before the entire cycle repeats:

$$M = \frac{1}{m} \sum_{i=1}^{m} x_i = \frac{1}{m} \sum_{i=1}^{m} i = \frac{m + 1}{2}$$

$$\sigma^2 = \frac{1}{m} \sum_{i=1}^{m} x_i^2 - M^2 = \frac{1}{m} \sum_{i=1}^{m} i^2 - \frac{(m + 1)^2}{4}$$

$$= \frac{(m + 1)(m - 1)}{12}$$

Normalizing to the $(0, 1)$ interval and assuming $m \pm 1 \approx m$, $(m + 1)(m - 1) \approx m^2$, for $m \gg 1$, $M_0 \approx 1/2$, and $\sigma_0^2 \approx 1/12$.

The *frequency test* is an empirical means of testing fractional period of nonperiodic sequences for uniformity. Numbers are produced on the $(0, 1)$ interval, divided into k subintervals, and a count of the number of numbers falling into each cell is recorded. Statistical significance is determined by a chi-square statistic with $k - 1$ degrees of freedom $(N/k > 5)$,

$$\chi^2_{freq} = \frac{k}{N} \sum_{i=1}^{k} \left(w_i - \frac{N}{k} \right)^2$$

w_i = empirical frequency count and N/k = theoretical frequency. Figure 4-1 shows plots of the frequency test applied to a "good" and "bad" RNG.

The *serial test* empirically determines serial association between the digits of successive numbers of the RNG sequence. Each digit is selected by normalizing to the $(0, k)$ interval or by selecting a portion of each random number and then scaling to $(0, k)$. The test is usually applied to pairs of digits and is an extension of the frequency test for 2-space. Therefore, k^2 cells are checked for uniformity using

$$\chi^2_{ser} = (k^2/N) \sum_{i,j} (w_{i,j} - N/k^2)^2,$$

and once again statistical significance is valid if $N/k^2 > 5$. Good [6] claims that $\chi^2_{ser} - \chi^2_{freq}$ has asymptotically a chi-square distribution with $k^2 - k$ degrees of freedom as $N \to \infty$. The exact distribution is not known, but others [7, 8] have merely extended the serial test to a two-dimensional frequency test with $k^2 - 1$ degrees of freedom. The two-dimensional analog will be assumed here. Figure 4-2 demonstrates a "good" and "bad" serial test result for the leading digits of a RNG with $k = 10$.

The *gap test* is performed by counting the length of gaps between successive like digits formed by normalizing the sequence to $(0, k)$. The gap length L is geometrically distributed:

$$P_L = (1 - 1/k)(1/k)^L; \quad L = 0, 1, \ldots.$$

$$X^2_{a-1} = \sum_{i=1}^{a} (g_i - NP_i)^2/(NP_i)$$

where a is chosen so that $p_a \approx 0$. N = sample size; g_i = empirical gap of length i, count. For $k = 9$, $a = 22$, $\chi^2_{21} = \sum_{i=1}^{22}[(g_i - N_{N_i})^2/(NP_i)]$. Figure 4-3 shows plots of the gap test applied to various digits as indicated.

Yule's test (5-digit sum) consists of taking the sum of 5-decimal digits from normalized RNG numbers and comparing it with the theoretically expected values.

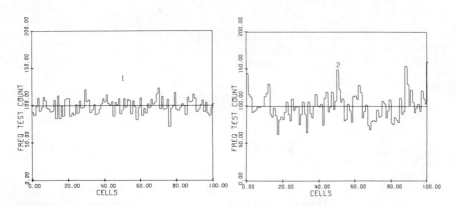

Figure 4-1. Frequency test results for (*a*) a "good" RNG and (*b*) a "bad" RNG.

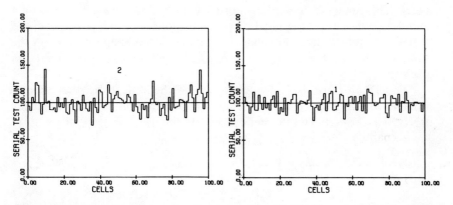

Figure 4-2. Serial test results for (*a*) a "bad" RNG and (*b*) a "good" RNG.

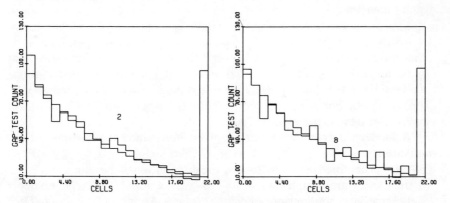

Figure 4-3. Gap test results for digits (*a*) 2 and (*b*) 8. HD2 generator.

$$P_i = 10^{-5} \sum_{r=0}^{5} (-1)^r \binom{5}{r} \binom{i - 10r + 4}{4}$$

$$\chi^2_{44} = \sum_{i=1}^{45} \frac{(y_i - NP_i)^2}{NP_i}$$

because

$$\max_i \left\{ \sum_{j=1}^{5} i \right\} = \sum_{j=1}^{5} 9 = 45$$

Yule's test detects "patchiness" in sequences by considering groups of 5 digits, but it could be extended to more than 5 digits. However, the central limit theorem comes into play for about 8 digits. The sum of $n > 8$ digits from any distribution will be approximately normally distributed in the mean. This tendency toward the normal distribution is clear in Figures 4-4 and 4-5. Yule's test is applied to the first 5 digits of each RNG number tested. Therefore, five chi-square values, corresponding to a test on the first, second, third, etc., digit of groups of five numbers, are obtained in each test.

The D^2 *test* consists of comparing the theoretical distribution of a random line in 2-space with an empirical distribution obtained from the RNG being tested. The theoretical distribution [9]

$$p(d) = \begin{cases} \pi d^2 - 8/3d^3 + 1/2d^4 & 0 \le d \le 1 \\ 1/3 + (\pi - 2)d^2 + 4(d^2 - 1)^{3/2} - 1/2d^4 - 4d^2 \sec^{-1} d & 1 < d \le \sqrt{2} \end{cases}$$

is compared with empirical values at discrete intervals by using a chi-square measure.

This test seeks to determine the suitability of an RNG for Monte Carlo evaluation of integrals (area) [10]. Accurate Monte Carlo integration over a plane requires uniform sampling of points from the plane. Clustering and/or sparseness of points in 2-space will show up as deviations in the distribution of a random line in 2-space since the line is determined by two endpoints. Thus, endpoint pairs are selected "randomly" when points are uniformly spaced throughout the planar region of interest.

A modern derivation of $p(d)$ is given here since the last published derivation was given in 1891 [11]. The distribution of $d^2 = z = (x_1 - x_2)^2 + (x_3 - x_4)^2$ is obtained in part by finding the distribution of the sum, $z = x + y$, where x and y are sampled according to $(x_1 - x_2)^2$ and $(x_3 - x_4)^2$, respectively; that is, x and y have density $(1/\sqrt{x} - 1)$ and $(1/\sqrt{y} - 1)$. The distribution of z, $H(z)$, is found by integration over the unit square.

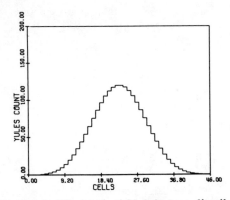

Figure 4-4. Theoretical Yule's test distribution.

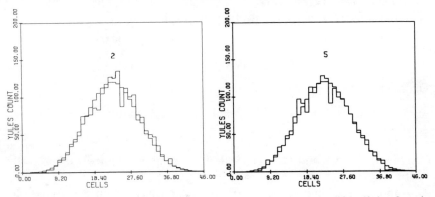

Figure 4-5. Yule's test result for (*a*) second digit and (*b*) fifth digit of each random number. HD2 generator.

CASE 1: $0 \leq z \leq 1$

$$H(z) = \int_{x=0}^{z} \int_{y=0}^{z-x} \left(\frac{1}{\sqrt{y}} - 1 \right) \left(\frac{1}{\sqrt{x}} - 1 \right) dx \, dy$$

$$= \int_{0}^{z} 2 \sqrt{\frac{z-x}{x}} \, dx - \int_{0}^{z} 2\sqrt{z-x} \, dx$$

$$- \int_{0}^{z} \frac{z-x}{\sqrt{x}} \, dx + \int_{0}^{z} (z-x) \, dx$$

By substitution of $x = z \cos^2 \theta$ in the first integral, $u = z - x$ in the second integral, $x = z \cos^2 \theta$ in the third integral, and combining results yields

$$H(z) = \pi z - 8/3z^{3/2} + z^2/2 \quad 0 \leq z < 1$$

or, for $z = d^2$, $P(d) = \pi d^2 - 8/3d^3 + d^4/2 \quad 0 \leq d < 1.$

CASE 2: $1 < z \le 2$

$$H(z) \;=\; \int_0^1 \int_0^{z-1} \left(\frac{1}{\sqrt{y}} - 1 \right) \left(\frac{1}{\sqrt{x}} - 1 \right) \, dx \, dy$$

$$+ \int_{z-1}^1 \int_0^{z-x} \left(\frac{1}{\sqrt{y}} - 1 \right) \left(\frac{1}{\sqrt{x}} - 1 \right) \, dy \, dx$$

$$= \; 2\sqrt{z - 1} - z - 1$$

$$+ \int_{z-1}^1 \left[\; \sqrt{\frac{z - x}{x}} - \frac{2 - x}{\sqrt{x}} + (z - x) \right] \, dx$$

Let $x = z \sin^2 \theta$, $\theta_L = \sin^{-1} \sqrt{(z - 1)/z}$, $\theta_u = \sin^{-1} \sqrt{1/z}$, and note that $\theta_L + \theta_u = \pi/2$ and $\theta_L = \sec^{-1} \sqrt{z}$.

After a horrendous amount of algebra,

$$H(z) = 1/3 + (\pi - 2)z + (2 + 2z)\sqrt{z - 1} + 2/3(z - 1)^{3/2}$$

$$- \; z^2/2 - 4z \sec^{-1} \sqrt{z}; \quad 1 \le z \le 2.$$

Or, after setting $z = d^2$ and further simplification,

$$P(d) = 1/3 + (\pi - 2)d^2 + 4\sqrt{d^2 - 1} + 8/3(d^2 - 1)^{3/2}$$

$$- \; d^4/2 - 4d^2 \sec^{-1} d; \quad 1 < d \le \sqrt{2}.$$

Combining Case 1 and Case 2 yields the "rainbow-shaped" theoretical distribution shown in Figure 4-6.

Approximate analytic *autocorrelation tests* have been performed on full period congruential generators [12, 13]. Fractional period sequences and other RNGs must be tested empirically by "correlograms" calculated from N-length sequences, $\{X_i\}$, $R(t) = (1/N)\Sigma_{i=1}^N X_i X_{i+t'}$, and the normalized autocorrelation function $R_{xx}(t) = R(t)/R(0)$. Greenberger [12] has shown that for HD2, $|R_{xx}(t)| \approx 1/\lambda + (12/p)(s/\lambda^2 - \lambda/4)$, where λ = HD2 RNG multiplier, p = period of HD2, and s = weak function of λ^3, leading to the conclusion that $\lambda \approx \sqrt{p}$ yields minimal correlation. Whittlesey [14] gives results of empirical autocorrelation tests using

$$\max_{0 < t \le 50} \; | \, R_{xx}(t) \, |$$

test statistic:

"for $N > 75$, the large sample (Gaussian) approximations can be used." Having determined that $R_{xx}(1)$, $R_{xx}(2)$, ..., $R_{xx}(50)$ are independent and approximately Gaussian distributed (mean approximately equal to 0, variance approximately equal to $1/N$), one can then readily obtain histograms for the distribution of max $| \, R_{xx}(1)$, $R_{xx}(2)$, ..., $R_{xx}(50) |$ by taking successive (two-sided) areas from this Gaussian probability density function, raising each of these areas to the 50th power, and then subtracting to obtain; (prob $(a_i < \max | R_{xx}(t) | \le a_{i+1})$.

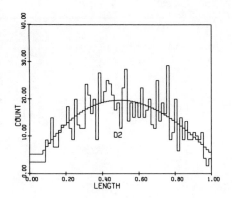

Figure 4-6. D^2 test result. HD2 generator.

Hence, $R_{xx}(t)$, $t = 1, 2, \ldots, 50$ is computed and the maximum tested. If

$$0.03 \leq \max_t \mid R_{xx}(t) \mid \leq 0.08, \quad N > 75$$

then the sequence is said to be uncorrelated. Figure 4-7 gives plots of $R_{xx}(t)$, $t = 1, 2, \ldots, 50$, for "good" and "bad" RNGs.

The *conditional bit test* [15] is performed to check independence of bits within each number produced by an RNG. Each number is treated as a binary integer and decomposed bit by bit. The conditional probability, $\text{Prob}(b_j = 1 \mid \text{first } j - 1 \text{ bits}; j = 1, 2, \ldots, n)$ is computed by counting the occurrence of 1s in each bit position, b_j, of n-bit integers, conditional on the previous $j - 1$ bits. A table is constructed with bit counts:

$$\text{Prob}(b_1 = 1)$$

$$\text{Prob}(b_2 = 1 \mid 0) \quad \text{Prob}(b_2 = 1 \mid 1)$$

$$\text{Prob}(b_3 = 1 \mid 00) \quad \text{Prob}(b_3 = 1 \mid 01) \quad \text{Prob}(b_3 = 1 \mid 10) \quad \text{Prob}(b_3 = 1 \mid 11)$$

$$\vdots$$

Chi-square tests of $1, 3, 7, \ldots, 2^n - 1$ degrees of freedom are performed on empirical counts which should agree with the expected result of 1/2.

The distribution of bits within each number of a sequence may be important in applications which use bits to construct other distributions. For example, the sum of n bits may be used to approximate the normal distribution [17]. Also, simulation of Markov processes which rely on conditional transition probabilities can be influenced by dependence of bits within each number of an RNG sequence. Other tests have been performed on RNG sequences but are not discussed here [16].

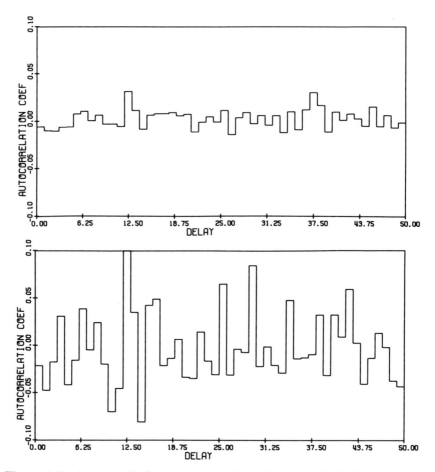

Figure 4-7. Autocorrelation test result for (*a*) a "good" RNG and (*b*) a "bad" RNG.

4-3 High-dimensional Tests

The distribution of random *n*-tuples (points in *n*-space) should be uniform for truly random sequences, but according to the *n*-cube theorem, $m^n - m$ vacant cells will appear in an *n*-cube. This anomaly may not necessarily be bad if low resolution suffices for a particular application, and if vacant cells (hence, filled cells) are uniformily scattered throughout the *n*-cube. Thus, empirical tests are made to determine practical *n*-space uniformity.

A problem in devising *n*-space tests is that according to the central limit theorem, the sum of random variates approximates a normal distribution regardless of the underlying distribution of the summed variates. For example, the D^2 test might be extended to a D^n test and random lines in *n*-space generated. However, the distribution of a random line in 8-space looks bell-shaped, and the approximation to the normal distribution is very close. Thus, the RNG used to generate random lines in a D^n test may or may not be uniformly distributed (this still may be a good test for independence). Empirical tests in high-dimensional space should be designed with this smoothing effect in mind.

A *runs test* consists of computing lengths of runs-up and runs-down and also counting the number of runs occurring in a sequence. An approximation for runs-up length can be made for the HD1 RNG as follows. $X_k \equiv A^k X_0 + c[(A^k - 1)/(A - 1)]$ (mod m). Thus, $A^k X_0 + c[(a^k - 1)/(A - 1)] \approx m$; $c > 0$ for a run-up of k numbers. The runs-up of length k will be distributed as a function of a random variable,

$$k(X_0) = \frac{\log [(A - 1)m + c]/c}{\log A}$$

assuming X_0 to be a uniformly distributed starting value for each run-up.

$$P(k) = \left[\frac{(A - 1)m + c}{(m - 1)(A - 1)} \log A \right] \exp (- k \log A)$$

The expected value is

$$\bar{k} = \log \left[\frac{(A - 1)(m - 1) + c}{c} \log A \right] \qquad c > 0$$

This approximation becomes worse as A approaches m, but note that $\bar{k} \,|\, _{A=m} \approx 2; c << m$.

For continuous uniform variates, the run length is exponentially distributed with mean of 1. There is no confusion here because $\bar{k}$ represents the number of numbers in a run-up. Hence, run length is construed to be the number of consecutive plus or minus signs obtained by replacing a sequence $X_1, X_2, \ldots$ with the sign of $(X_{i+1} - X_i)$. For example, 1, 5, 19, 15, 18, $\ldots$ yields $+, +, -, +, \ldots$. Figure 4-8 shows the rapid drop to the desired runs-up performance of HD1 as the multiplier, and A is increased.

Empirical verification of runs-up and runs-down consists of counting the number of runs in a sequence of length N and the runs above or below the mean, and comparing the distribution of run lengths with the theoretical [18], $E(r) = (1/3)(2N - 1)$; var $(r) = (16N - 29)/90$; $E(r_m) = N/2 + 1$; where r = number of runs in sequence, r_p = number of runs of length p, and r_m = number of runs above or below the mean.

$$E(r_p) = \frac{2N(p^2 + 3p + 1) - 2 (p^3 + 3p^2 - p - 4)}{(p + 3)!}$$

An empirical determination of runs of length p is made and tested for chi-square significance, as shown in Figure 4-9.

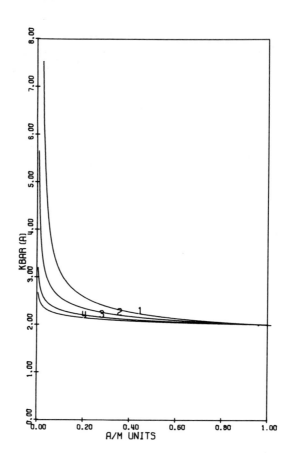

Figure 4-8. Average runs-up versus multiplier A.

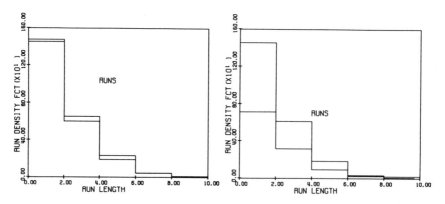

Figure 4-9. Runs test results for (a) a "good" RNG, and (b) a "bad" RNG.

The *sum-of-n test* can be thought of as an extension of Yule's test, and it checks an RNG sequence for uniformity and independence [19]. The sum of n numbers from an RNG is computed N times, and a frequency count is recorded by dividing the sum range into k cells. The expected count P_i for the ith cell is distributed as

$$F(P_i) - \frac{1}{n!} \sum_{j=0}^{[P_i]} (-1)^j \binom{n}{j} (P_i - j)^n$$

A chi-square test with $k - 1$ degrees of freedom determines significance. Figure 4-10 reveals a very close fit to the theoretical cumulative distribution for $n = 3, 5$.

The *maximum/minimum-of-n test* detects extreme-value nonuniformity. If $X_1, X_2, \ldots, X_n$ are normalized $(0, 1)$ RNG numbers assumed to be uniform, then

$$W = \max_{1 \le i \le n} \{X_i\}$$

is distributed, $F(a) = a^n = \text{Prob}(W < a)$ for $0 < a < 1$. Then, the statistic W^n has distribution, $\text{Prob}(W^n < a) = F(a^{1/n}) = a$, and W^n is uniformly distributed. W^n is computed N times, and a frequency test is applied to the number of numbers occurring in each of 100 equal cells on the unit interval.

For the minimum extreme,

$$W^n = 1 - \left(1 - \min_{1 \le i \le n} \{X_i\} \right)^n$$

is the test statistic and is also uniform. Figure 4-11 shows results for max of 10 and $N = 10,000$.

Frequency analysis of RNG sequences has been done in closed form for full period Lehmer (HD1) and Tausworthe (FSR) RNGs. Coveyou and MacPherson [19] performed a finite Fourier transform (FFT) analysis on HD1, which revealed wave properties discussed in Chapter 2. Tausworthe performed the analog of FFT using Rademacher-Walsh functions

$$\phi(s, x) = 2^{-n/2} (-1)^{s_1 x_1 + s_2 x_2 + \cdots + s_n x_n}$$

$$s_i, x_i = 0, 1$$

and the orthogonal expressions

$$G(s) = 2^{-n/2} \Sigma_x g(x) \phi(s, x)$$

$$g(x) = 2^{n/2} \Sigma_s G(s) \phi(s, x)$$

The pattern function $g(x)$ determines a matched pattern in the bits $\{a_i\}$:

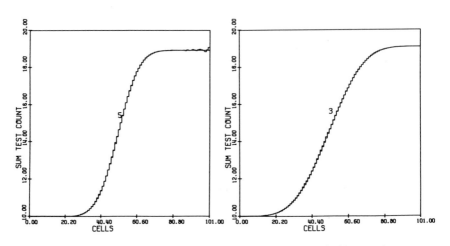

Figure 4-10. Sum-of-n test for (a) $n = 5$ and (b) $n = 3$.

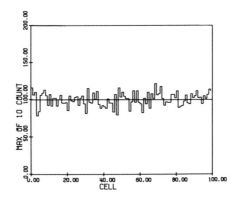

Figure 4-11. Max-of-10 test result. HD2 generator.

$$g(x) = \begin{cases} -1 & (x_1, x_2, \ldots, x_n) = (a_i, a_{i+1}, \ldots, a_{i+n-1}) \\ +1 & \text{otherwise} \end{cases}$$

$$= 1 - 2\delta_{a_i s_1} \delta_{a_{i+1} s_2}, \ldots, \delta_{a_{i+n-1} s_n}$$

where

$$\delta_{rs} = \begin{cases} 1 & r = s \\ 0 & r \neq s \end{cases}$$

An empirical test [20, 21, 22] yields practical results for any sequence by applying tests to the periodogram obtained by estimating the spectrum as follows: a random sample, $\{X_j, j = 1, 2, \ldots, N\}$ is transformed

$$a_n = \frac{1}{N} \sum_{j=0}^{N-1} X_j \exp\left(-2\pi ij \frac{n}{N}\right) \qquad n = 0, 1, \ldots, N-1$$

to obtain the normalized cumulative periodogram points

$$P_n = \sum_{r=1}^{N} \frac{|a_n|^2}{\sum_{r=1}^{M+1} |a_n|^2} \qquad n = 1, 2, \ldots, M+1$$

$$M = \begin{cases} N/2 - 2 & N \text{ even} \\ (N-3)/2 & N \text{ odd} \end{cases}$$

And the normalized *median-spectrum test* statistic

$$U = (s - \tfrac{1}{2})\sqrt{12M} \qquad s = \frac{1}{M} \sum_{n=0}^{M} P_n$$

which has a unit normal distribution to be tested for uniformity of frequency components of the sequence. If an RNG fails this test, a periodicity must exist in the sequence, and the desired "flat" spectrum is not realized.

The *Kolmogorov-Smirnov test* statistic may be applied for goodness of fit:

$$Ks^+ = \max_{1 \le n \le m} \frac{P_n - n}{m+1}$$

$$Ks^- = \min_{1 \le n \le m} \frac{n}{M + 1 - P_n}$$

$$Ks = \max\{Ks^+, Ks^-\}$$

To test for a *constant spectrum* (stationarity), a modified Bartletts test for variance heterogeneity may be applied:

$$T_i^2 = \frac{\sum_{n=(i-1)\sqrt{v}+1}^{i\sqrt{v}} P_n}{2\pi} \qquad i = 1, 2, \ldots, k$$

where the M values of P_n are divided into K contiguous groups of size v and k is the largest integer such that $1 \le kv \le M$. The test statistic is

$$H(k) = \frac{21 \log\left(\sum_{i=1}^{k} T_i^2/21\right) - \sum_{i=1}^{k} 2v \log[T_i^2/(2v)]}{(6v-2)/(6v-3)}$$

If $2v$ is greater than 5, then $H(k)$ is approximately chi-square with $k - 1$ degrees of freedom. A problem in using this statistic is the arbitrary choice of k. These have been selected as $k = 50$, $N = 512$, and $M = 254$ in Figure 4-12.

In addition to statistical tests based on theoretical distributions, a test relating to the use to be made of an RNG should be performed. Thus, the following simulation is used as an applied test under "battle conditions."

A *scattering experiment* [22] is simulated as shown in Figure 4-13. An entering beam is deflected by collisions with atoms in a spherical medium. The simulation is carried out by sampling random points from the sphere's surface, rejecting the subsequent solid angles according to a scatter cross-sectional curve, and using a suitable solid angle to direct the beam toward a second atom, etc. Hence, the simulation is done as two sampling experiments: (1) A solid scattering angle is selected by sampling from a distribution uniform over the surface of a unit sphere, $u^2 + v^2 + w^2 = 1$, $w = 2R_1 - 1$, $\phi = \pi(2R_2 - 1)$, $0 \le R_1, R_2 \le 1$, $u = \sqrt{1 - w^2} \cos \phi$, $v = \sqrt{1 - w^2} \sin \phi$. (2) Then a rejection of the angles from (1) is applied to weight according to the cross-sectional curve $\sigma(\theta)$. For this experiment, $\sigma(\theta) = \pi(1 - \theta/\pi)$, $0 \le \theta \le \pi$.

An empirical curve is obtained by running this simulation for 36 subdivisions of θ (5 degrees each). The count of numbers in each subdivision $S(\theta_i)$ is compared with the expected count for N angles using a chi-square test:

$$X_s^2 = \frac{\pi}{36N} \sum_{i=1}^{36} \frac{\left[S(\theta_i) - \dfrac{N\sigma(\theta_i)}{\pi} \right]^2}{\sigma(\theta_i)}$$

4-4 Results of Tests

Appendix A shows test results applied to HD2 with multiplier $a = 14^{29}$, mod $2^{31} - 1$; HD1 with multiplier $\lambda = 2{,}719{,}982{,}433$; composite bank of HD2 RNGs with multipliers 7^5, 14^{29}, 31^5, 45^5, 11^3, 39^{13}, 53^{13}, 88^{13}, 95^{13}, 14^{17}, 22^{17}, 44^{17}, 45^{17}, 53^{17}, 51^{17}, 62^{17}, respectively; Kendall's FSR algorithm with $x^{31} + x^{13} + 1$; and the GFSR algorithm using $x^{98} + x^{27} + 1$ and delayed columns equaling $98 \cdot (100)$ bits. Each RNG is 31 bits in length except GFSR which has a 15-bit word size. The tests were performed on samples of 10,000 numbers.

In each case of a failure at the 5 percent level of significance, an * is included. Only those tests which resulted in a failure at the 5 percent level will be discussed. All RNGs failed both the sum-of-n test and the "constant spectrum" part of the FFT test. However, the composite HD2 RNG

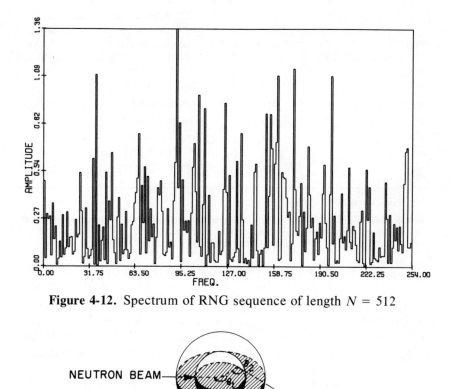

Figure 4-12. Spectrum of RNG sequence of length $N = 512$

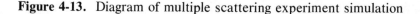

MULTIPLE SCATTERING EXPERIMENT

Figure 4-13. Diagram of multiple scattering experiment simulation

performed best on the sum-of-n test, indicating that this test is detecting independence of successive numbers. The FFT test would indicate a "moving spectrum" in each RNG, but doubts as to the appropriateness of this test seem to be verified here. In most cases, failures were removed by taking larger samples. For the GFSR, Yule's test was satisfied by taking 50,000 numbers. In the cases where more than 15-bit accuracy is needed, "$===$" is used.

Failures persisted for minimum-of-n test, $n > 6$, GFSR. This agrees with theory for $n \le p/L = 6$, from Chapter 3. Hence, this test partially verifies theoretical predictions for this small-word-size RNG.

RNG	Speed Estimates for IBM 360/65	Generality	Statistical Tests Failed
HD2	28 M sec.	Machine dependent	6
HD1	29 M sec.	Machine dependent	7
16HD2	30 M sec.	Machine dependent	7
FSR	25 M sec.	Machine dependent	10
GFSR	*20 M sec.	Portable in higher level language	7
GFSR Microcoded	* 5 M sec.	Machine dependent	7 (simulated)

*based on Interdata 4 estimates

Source: P. D. Gross and W. H. Payne, "Comparison of shift register sequence and congruential pseudorandom number generators implemented on the SRU 1100, IBM 360, and CDC 6000 series computers." Washington State University, 4 (June, 1970).

Figure 4-14. Summary of RNG performance.

Apparently, these empirical tests fail to detect n-space nonuniformity predicted in Chapter 2. There are two possible explanations: (1) the integer numbers produced are normalized to floating-point numbers in (0, 1), and (2) the tests actually use only high-order or small groups of bits from each pseudorandom number.

The normalization process truncates lower-order bits and produces duplicate numbers within the sequence. The resolution is cut down, but not enough to ruin the approximation to a continuous variate. The duplicates aid in filling n-space as desired.

The tests often use high-order bits or require such low resolution that nonuniformity is undetectable. A frequency test on all m^n cells is impractical even on a very large storage, fast computer. So, from a practical standpoint, the theoretical predictions of nonuniformity are not severe limitations for the word sizes tested here.

Obviously, the ability of each test to determine nonrandomness is varied. Some tests are more powerful than others. If we define power as equal to the number of failures divided by the number of tests, these tests can be ordered from most powerful to least powerful: sum-of-n (0.68), correlation (0.40), FFT (0.33), scatter (0.20), Yules (0.12), runs (0.10), min/max (0.08), gap (0.04), conditional bit (0.03), and frequency, D^2, and serial (0.0).

In terms of the number of tests failed, HD2 failed 6 times, HD1 failed 7 times, 16 HD2 failed 7 times, FSR failed 10 times, and GFSR failed 10 times. Considering theoretical predictions of failure for GFSR in space ($n = 6$), one must conclude that the performance of GFSR with 7 failures compares favorably with the others considered here. In addition, GFSR is enhanced by speed, generality, and portability, as discussed in Chapter 3. The GFSR algorithm must be considered statistically equal to the other RNGs tested, but with superior speed and generality, as shown in Figure 4-14.

References

[1] Hull, T.E., and Dobell, A.R. Random number generators. *SIAM Review*, 4: 230-254 (1962).

[2] Kendall, M.G., and Smith, B.B. Randomness and random sampling numbers. *J. Roy. Statis. Coc.*, 101: 162-164 (1938).

[3] Chambers, R.P. Random number generation on digital computers. *IEEE Spectrum*, 48-56 (Feb. 1967).

[4] MacLaren, M.D., and Marsaglia, G. Uniform random number generators. *J. ACM.*, 12: 83-89 (1965).

[5] Lewis, P.A.W., Goodman, A.S., and Miller, J.M. A pseudorandom number generator for the system/360. Unpublished report.

[6] Good, K.J. On the serial test for random sequences. *Am. Math. Statis.*, 28: 262 (1957).

[7] MacLaren, M.D., and Marsaglia, G. Uniform random number generators. *J. ACM.*, 12: 83-89 (1965).

[8] Lewis, P.A.W., Goodman, A.S., and Miller, J.M. A pseudorandom number generator for the system/360. Unpublished.

[9] Tocher, K.D. "The Art of Simulation." London: English University Press, Ltd., 1963, p. 47.

[10] Gruenberger, F., and Mark, A.M. The d^2 test of random digits. *Math. Tables Other Aids Comp.*, 5 (1951).

[11] Wilson, B. "Integral Calculus." London: 1891, p. 390.

[12] Greenberger, M. An a priori determination of serial correlation in computer generated random numbers. *Math. Comp.*, 15: 383-389 (1961).

[13] Good, K.J. On the serial test for random sequences. *Am. Math. Statis.*, 28: 262 (1957).

[14] Whittlesey, J.R.B. A comparison of the correlational behavior of random number generators for the IBM 360. *Comm. ACM.*, 11: 641-644 (Sept. 1968). Copyright 1968, by *Association for Computing Machinery, Inc.*, reprinted by permission.

[15] Payne, W.H., and Lewis, T.G. Conditional bit sampling: accuracy and speed, in "Mathematical Software." New York: Academic Press, 1971.

[16] White, R.C. Experiments with digital computer simulations of pseudo-random noise generators. *IEEE Trans. Elect. Comp.* (*Short Notes*), 355-357 (June, 1967).

[17] Gorenstein, S. Testing a random number generator. *Comm. ACM.*, 10: 111-118 (Feb. 1967).

[18] MacLaren, M.D., and Marsaglia, G. Uniform random number generators. *J. ACM.*, 12: 83-89 (1965).

[19] Coveyou, R.R., and MacPherson, R.D. Fourier analysis of uniform random number generators. *J. ACM.*, 14: 100-119 (Jan. 1967).

[20] Cox, D.R., and Lewis, P.A.W. "The Statistical Analysis of Series of Events." New York: Barnes and Noble and London: Methueu, 1966.

[21] Lewis, P.A.W., Goodman, A.S., and Miller, J.N. A pseudo-random number generator for the system/360. Unpublished report.

[22] Tripard, G.E. The production of a well collimated neutron beam using the associated particle technique. Ph.D. thesis, University of British Columbia, 1967.

[23] Gross, P.D., and Payne, W.H. Comparison of shift register sequence and congruential pseudo-random number 6000 series computers. Washington State University. *Time Slicer* 4 (June 1970).

Part II
Nonuniform Distribution
Sampling

5

Inversion Technique

5-1 Introduction

Consider the following "ruin" problem. The U.S. Widget Company plans an ambitious advertising campaign that is estimated to have probability p of success and probability q of failure. If the plan succeeds, the company will increase its current worth $z = \$10$ million to $a = \$20$ million. The problem, however, is that the company must risk $1 million annually to "play" its advertising game. If the yearly plan succeeds, Widget will earn $2 million. The company president has made it perfectly clear to the board of directors that "we intend to use our strategy every year until Widget either goes broke or else becomes a $20 million company."

Young Smith, an upward-mobile executive, challenged the plan because Smith claims that Widget is not playing in a fair market unless the probability of success in each year is $p = 1/3$ and the probability of failure is $q = 2/3$. Indeed, if $p < 1/3$, then the company will be ruined eventually, and if $p = 1/3$, the company only stands a 50 percent chance of success. Smith suggests that the Widget computer facility construct a simulation model of the company and its market before proceeding with this dangerous strategy.

The object of Smith's simulation is to estimate p and q and then compute $p_z = $ the probability of ruin, that is, the probability of losing the original $z = \$10$ million.

The ruin probability p_z is bounded by upper and lower values, as given by the Gambler's Ruin program (Figure 5-1). In a general game where a gambler bets β units, to win a play of the game, which pays α units, the single-play probability of winning is p, the probability of losing is q, and that of tieing is r:

$$\lambda^z \frac{\lambda^{a-z-(\alpha-1)} - 1}{\lambda^{a-(\alpha-1)} - 1} \le p_z \le \lambda^{z-(\beta-1)} \frac{\lambda^{a-z} - 1}{\lambda^{a-(\beta-1)} - 1}$$

where $z = $ original wealth of player

$p = $ the probability of winning

$q = $ the probability of losing

$r = $ the probability of tieing

```
C        PURPOSE:  TO CAL. THE PROB. OF THE GAMBLER'S RUIN.
C        REFERENCE:  RICHARD A. EPSTEIN,THE THEORY OF GAMBLING
C        AND STATISTICAL LOGIC,ACADEMIC PRESS INC.,NEW YORK,1967
C        METHOD:  DETERMINE THE UPPER AND LOWER BOUNDS OF THE
C        PROB. OF RUIN BY USING THE INEQUALITIES AS STATED IN
C        THE REFERENCE.  IN ORDER TO CAL. THESE BOUNDS IT IS
C        NECESSARY TO DETERMINE THE ROOT(LAMBDA) OF A PARTICULAR
C        EXPONENTIAL EQUATION.  THIS EXPONENTIAL EQ. IS ALSO
C        STATED IN THE REFERENCE.
C        PROGRAMMER:  WARREN V. CAMP,U.S.L.,19 SEPT. 1974
         IMPLICIT REAL*8(A-L,O-Z)
         WRITE(2,101)
101      FORMAT(' ENTER THE FOLLOWING DATA IN FLOATING PT FORMAT')
         WRITE(2,102)
102      FORMAT(' PROP. OF A WIN, EXAMPLE 0.5')
         READ(1,103) P
103      FORMAT(F10.5)
         WRITE(2,104)
104      FORMAT(' PROB. OF A LOSS, EXAMPLE 0.35')
         READ(1,103) Q
         WRITE(2,105)
105      FORMAT(' AMOUNT WON UPON EACH PLAY, EXAMPLE 32.0')
         READ(1,106) X
106      FORMAT(F14.2)
         WRITE(2,107)
107      FORMAT(' AMOUNT BET AT EACH PLAY, EXAMPLE 10.00')
         READ(1,106) B
         WRITE(2,108)
108      FORMAT(' INITIAL WEALTH, EXAMPLE 1000.0')
         READ(1,106) Z
         WRITE(2,109)
109      FORMAT(' FINAL WEALTH TO STOP PLAY OR ELSE RUIN, EXAMPLE 20000.0')
         READ(1,106) A
         IF(A.LT.(Z+B)) GO TO 12
         SAVA = A
         SAVZ = Z
C        CAL. THE MEAN EXPECTATION FOR A SINGLE PLAY.
         EXPCT = X * P - B * Q
C        IF THE GAME IS UNFAIR THEN GO TO 1 ELSE CAL.
C        THE PROB OF RUIN USING THE FOLLOWING EQS.
         IF(EXPCT.NE.0.0) GO TO 1
         LAMBDA = 1.
         PZMIN = (A-Z-(X-1.))/(A-(X-1.))
         PZMAX = (A-Z)/(A-(B-1.))
         GO TO 11
1        DIVSOR = X
         IF(X.GT.B) DIVSOR = B
         X = X / DIVSOR
         B = B / DIVSOR
         Z = Z / DIVSOR
         A = A / DIVSOR
         EXPONT = X/B + 1.
```

Figure 5-1. Gambler's ruin program.

α = the amount to be won

β = the amount to be lost

a = the amount needed by player before stopping, and $a \geq z+\beta$

λ = root of the equation $p\lambda^{\alpha+\beta} - (1 - r)\lambda^{\beta} + q = 0$

If the player's wealth decreases to less than β, then the player has been

```
      R = 1. - P - Q
C     TEST FOR POSSIBLE OVERFLOW, IF SO STOP.
      TSTVAL = (B/X) * DLOG((B*(1.-R))/(P*(X+B)))
      IF(TSTVAL.GT.172.) GO TO 12
C     CAL. THE VALUE OF LAMBDA AT WHICH A
C     MINIMUM OCCURS FOR THE EXPONENTIAL EQ.
      EQMIN = ((B*(1.-P))/(P*(X+B)))**(B/X)
      IF(EQMIN.EQ.1.) GU TO 12
C     TEST FOR POSSIBLE OVERFLOW, IF SO STOP.
      TSTVAL = EXPONT * DLOG(EQMIN) + DLOG(P)
      IF(TSTVAL.GT.172.) GO TO 12
C     CAL. THE MINUMUM POINT OF THE EQUATION
      FNMIN = P*(EQMIN**EXPONT) - (1.-R) * EQMIN + Q
      WRITE(2,114) EXPCT
114   FORMAT('          MEAN EXPECTATION = ',F14.7)
C     IF MINIMUM LESS THAN 0, THEN ANOTHER ROOT
C     EXISTS OTHER THAN LAMBDA=1
      IF(FNMIN.LT.0.0) GO TO 2
      LAMBDA = EQMIN
      GU TO 8
C     IF THE MEAN EXPECTATION IS NEGATIVE, THEN
C     ROOT OF THE EQ. IS BETWEEN 1 & SOME UPPER
C     BOUND, ELSE THE ROOT IS BETWEEN 0 & 1.
2     IF(EXPCT.LT.0.0) GO TO 3
      X1 = 0.0
      X2 = EQMIN
      GO TO 5
C     DETERMINE AN UPPER BOUND FOR THE ROOT(LAMBDA).
3     X1 = EQMIN
      X2 = EQMIN
4     X2 = X2+2.0
C     TEST FOR POSSIBLE OVERFLOW, IF SO STOP.
      TSTVAL = (EXPONT*DLOG(X2)) + DLOG(P)
      IF(TSTVAL.GT.172.) GO TO 12
      FB = DEXP(TSTVAL) - (1.-R)*X2 + Q
      IF(FB.GT.0.0) GO TO 5
      IF(X2.GE.1000.) GU TO 12
      GU TO 4
C     THE MID-POINT METHOD IS USED TO DETERMINE THE
C     ROOT OF THE EXPONENTIAL EQ.
5     X3 = (X1+X2)/2.0
      FX3 = P*(X3**EXPONT)-(1.-R)*X3+Q
      IF(DABS(FX3).LT.0.00000001) GO TO 7
      FX1 = P*(X1**EXPONT)-(1.-R)*X1+Q
      IF((FX1*FX3).GE.0.0) GO TO 6
      X2=X3
      GO TO 5
6     FX2 = P*(X2**EXPONT)-(1.-R)*X2+Q
      IF((FX2*FX3).GE.0.0) GO TO 12
      X1 = X3
      GO TO 5
7     LAMBDA = X3**(1./B)
```

ruined and the game must stop. If this wealth builds to the set goal of $\$a$, then the player also stops because the goal has been reached. A program is given to compute p_z and $(1 - p_z)$. The game is fair only if

$$\alpha p - \beta q = 0$$

and we can see that Smith's statement is indeed true for $p = 1/3$, $q = 2/3$, and $\alpha = 2$, $\beta = 1$.

```
C       RESULTING FROM VERY LARGE EXPONENTS.
        TSTVAL = (A-Z) * DLOG(LAMBDA)
        IF(TSTVAL.GT.172.) GO TO 9
        IF(TSTVAL.LT.-170.) GO TO 10
C       CAL. THE LOWER BOUND OF THE RUIN PROB.
  8     PZMIN = ((LAMBDA**(A-Z-(X-1.))-1.)/(LAMBDA**(A-Z-(X-1.))
        A-LAMBDA**(-Z)))
C       CAL. THE UPPER BOUND OF THE RUIN PROB.
        PZMAX = ((LAMBDA**(A-Z)-1.)/(LAMBDA**(A-Z)-LAMBDA**(-Z+(B-1.))))
        GO TO 11
C       USE THESE EQS. WHEN "A" >>> "Z" & LAMBDA > 1.
  9     PZMIN = 1.
        PZMAX = 1.
        GO TO 11
C       USE THESE EQS. WHEN "A" >>> "Z" & LAMBDA < 1.
 10     PZMIN = LAMBDA ** Z
        PZMAX = PZMIN
C       CAL. THE PROB. OF SUCCESS.
 11     BARMAX = 1. - PZMIN
        BARMIN = 1. - PZMAX
C       CAL. THE GAMBLER'S EXPECTED GAIN.
        GMAX = SAVA * BARMAX - SAVZ
        GMIN = SAVA * BARMIN - SAVZ
        WRITE(2,110) LAMBDA
110     FORMAT(//,'         LAMBDA = ',F14.7)
        WRITE(2,111) PZMIN,PZMAX
111     FORMAT(' ',//,' ',F14.7,' ',' <=    P(Z)    <= ',F14.7)
        WRITE(2,112) BARMIN,BARMAX
112     FORMAT(' ',//,' ',F14.7,' <= (1-P(Z)) <= ',F14.7)
        WRITE(2,115) GMIN,GMAX
115     FORMAT(' ',//,' ',F14.3,' <= EXP GAIN <= ',F14.3)
        GO TO 999
 12     WRITE(2,113)
113     FORMAT(' *THIS METHOD CAN NOT BE USED TO CAL THE RUIN PROB*')
999     STOP
        END
```

Figure 5-1 (continued)

The Widget computer facility is using a simulation model to estimate probabilities p and q. They use a common definition of probability that permeates simulation modeling. Let N trials be performed by the simulation. These trials will be partitioned into three subsets: the subset of "successes," the subset of "failures," and the subset of "ties." The sizes of each subset are denoted by the following:

$$n_p = \text{number of successes}$$

$$n_q = \text{number of failures}$$

$$n_r = \text{number of ties}$$

The frequency of occurrence of a success, a failure, or a tie is estimated by computing these ratios:

$$f_p = n_p/N$$

$$f_q = n_q/N$$

$$f_r = n_r/N$$

```
ENTER THE FØLLØWING DATA IN FLØATING PT FØRMAT
PRØB. ØF A WIN, EXAMPLE 0.5
*0.3333
PRØB. ØF A LØSS, EXAMPLE 0.35
*0.6667
AMØUNT WØN UPØN EACH PLAY, EXAMPLE 32.0
*2000000.0
AMØUNT BET AT EACH PLAY, EXAMPLE 10.00
*1000000.0
INITIAL WEALTH, EXAMPLE 1000.0
*10000000.0
FINAL WEALTH TØ STØP PLAY ØR ELSE RUIN, EXAMPLE 20000.0
*20000000.0



        MEAN EXPECTATIØN =    -100.0000000

        LAMBDA =     1.0000958

        0.4739111 <=   P(Z)    <=      0.5002394

        0.4997606 <= (1-P(Z)) <=      0.5260889

        -4788.776 <= EXP GAIN <=      521778.982

143.
```

In simulation, we say that the frequency of occurrence of an event is an estimate of the probability for the given event. Thus, we use f_p as an estimate of p, f_q as an estimate of q, and f_r as an estimate of r.

A simulation or computer definition of probability is one that seeks the limiting value of f_p, say, for increasing the number of trials.

$$p = \lim_{N \to \infty} f_p$$

$$q = \lim_{N \to \infty} f_q$$

$$r = \lim_{N \to \infty} f_r$$

Unfortunately, computer models seldom run long enough to replace N with infinity in our definition. Thus we are never able to compute a probability by simulation. How close can we come by increasing N? The effects of increasing N are shown by the famous limit theorem.

$$\left| f_p - p \right| \leq X_c \sqrt{\frac{\text{variance}}{N}}$$

where $\quad X_c =$ critical value obtained from the student t distribution

variance = mean square error of simulation trials

The significant feature of the limit formula is its form. It shows that increases in N are not directly related to a decrease in the error. Let error = $|f_p - p|$ and $X_c \sqrt{\text{variance}} = $ constant (roughly). Then

$$\text{Error} = \frac{\text{constant}}{\sqrt{N}}$$

That is to say, increasing N by tenfold decreases the error by roughly one-third. Thus, we see that probability estimates are difficult to obtain.

In our hypothetical Widget problem, we have assumed a single probability distribution. It is likely that most models will deal with a wide array of probabilities corresponding with an array of outcomes. We may extend the definitions by assuming that the N trials are partitioned into a collection of subsets. Let the subsets be numbered 1, 2, 3, ..., k. Then we have these estimates:

$$f_1 = n_1/N$$
$$f_2 = n_2/N$$
$$\vdots$$
$$f_k = n_k/N$$

The collection of estimates $\{f_1, f_2, \ldots, f_k\}$ is called a *histogram*. The limiting values of f_i for $i = 1, 2, \ldots, k$ result in a discrete probability density function:

$$\lim_{N \to \infty} \{f_1, f_2, \ldots, f_k\} = \{p_1, p_2, p_3, \ldots, p_k\}$$

Therefore

$$\text{pdf} = \{p_1, p_2, \ldots, p_k\}$$

is the discrete probability density function (pdf) corresponding to the histogram of the array of events.

In addition, if k is allowed to increase without bound and n_i is allowed to shrink suitably for all i, then we can define a continuous probability density function as the function $p(x)$ defined over an interval (a,b).

$$\lim_{\substack{k \to \infty \\ n_i \to 0}} (\text{pdf}) = p(x) \qquad a \leq x \leq b$$

The distributions in this book are theoretically of both types: discrete and continuous. The processing of "probabilities" is essentially discrete because digital computers are discrete mathematics processors. Hence, in all cases, we will be dealing with discrete probabilities.

Complex processes not amenable to analytic solution can be economically solved by using computer simulation and modeling which, in turn, rely on distribution sampling. For example, simulations of traffic networks, biological processes, erratic behavior of neutrons, or computers themselves are sufficiently difficult to prevent analytic description.

Theoretical results of mathematical probability may be verified by sampling from distributions upon which theory is founded. Simulation provides an economical check on analytic results.

In Part II, we concentrate on methods for obtaining event i with probability p_i, as defined before. The event is in turn used to activate a simulation model.

5-2 History

Sampling was first used to determine the t distribution by "student" [1, 2]. The population was selected from a series of 3000 pairs of measurements obtained by index finger measurements of criminals written on cards "sampled" by drawing from a shuffled deck. Consecutive sets of four cards were drawn to calculate the mean, standard deviation, and correlation.

Bispham [1, 2] apparently was the first to sample from an arbitrary theoretical population. A population of 30 counters was drawn without replacement from urns. The individual counters were handled on the order of 0.5 million times and "would not have been possible without the generous help of school children . . . ."

The first published table of random digits, due to Tippett [1, 2], were sampled from 1000 small cards placed in a bag. After its digit was recorded, each card was replaced and the cards in the bag were mixed well. These numbers were converted into uniform random numbers by means of a key.

Other tables were produced from observing a spinning disc (1938), from using Selective Service lists (1942), and from the last digits of logarithms (1947). Finally, the Rand Corporation (1951) produced 1 million random digits by using a mechanical device. This table has found use in war "games" played by the American military.

Pearson [1, 2] suggests the "frequency class" method to select (almost) normal deviates. A frequency table approximation to the normal cdf (cumulative distribution function) is given by

$$p_k = \int_{x_k-0.05}^{x_k+0.05} \frac{1}{\sqrt{2\pi}} e^{-t^2/2} \, dt \qquad k = 0, \pm 1, \ldots, \pm 34$$

A random p_k is selected by using four Tippett digits to form a random decimal between zero and 0.9999. The X_k corresponding to p_k yields a normal number. Later, "frequency class" transformations on Tippett's numbers were laboriously performed and made available by Mahalanobis [1].

An extended "frequency class" method called "inverse probability integral" was used by Wold [1] to transform random numbers to *normal deviates*.

The inverse

$$p = \int_{-\infty}^{X} \frac{1}{\sqrt{2\pi}} \, e^{-t^2/2} \, dt$$

was effected by looking it up in a table or by selecting (from a well-stirred bag) a ball with an X written on it.

Sampling to solve integral and differential equations has been termed the "Monte Carlo technique" since the mid-1940s. Monte Carlo integration has been shown to be competitive with classical summation schemes for problems of dimension higher than 3. Monte Carlo techniques will be distinguished from distribution sampling discussed here.

Automatic Random Numbers

Tabulating machines were first used in sampling experiments in 1928 to reduce computational labor. Later, in 1937, punched cards were used to "randomize" digits [1].

A uniformly distributed number obtained artificially in a computer is called a *pseudorandom number* or simply a random number (see Part 1). Sampling from arbitrary distributions is performed through a transformation on the random numbers; therefore, swift generation of truly uniform random numbers is basic to automatic sampling from distributions. Early random number generators were either independent physical devices or sometimes devices connected to digital computers. Arithmetical processes easily implemented as computer programs have proved far more successful.

The "midsquare" method [1, 2] was mentioned by Mauchly (1949), Hammer (1951), and Forsythe (1951). A sequence of k-digit numbers is generated by selecting the middle k digits of the square of a previous k-digit number. The sequence may degenerate to a sequence of zeros, or it may repeat itself; thus, other methods were sought.

The "residue class" method of Lehmer (1949) has been widely successful. A sequence Z_i is generated:

$$Z_i = AZ_{i-1} \quad (\text{modulo } m)$$

The modulus m is prime, and A is a primitive root which generates a maximum sequence length (see Part I).

The "shift register" sequence method of Tausworthe, based on Galois field theory, also provides a fast procedure for generating pseudorandom numbers. Maximum-length sequences of random numbers are generated by performing shifts and logical operations on the previous number contained in the computer's working register (see Part I).

Modern Sampling

A q random number will be a number sampled from the q distribution on a computer, and the corresponding algorithm will be termed a *q generator*. A particular q generator is said to be good if it

(1) requires little computer time, i.e., is fast

(2) requires little memory space, i.e., is compact

(3) generates high-quality numbers, i.e., is accurate

(4) is easy to implement, i.e., is portable

A binary code can be divised to rate each generator according to criteria (1) through (4). For example, a generator said to be compact (2), and accurate (3), is rated 0110 = 6.

5-3 Inversion Sampling

A general technique for generating variates which obey a probability density function, pdf, is the *inverse transformation* method. This method utilizes the fact that the cumulative distribution function, cdf, corresponding to a pdf, can be inverted by using values selected from the uniform, $U(0, 1)$, distribution. $U(0, 1)$ means a random number in the interval $(0, 1)$.

METHOD Suppose variates X are to be sampled from pdf $f(X)$. Then,

$$\int_{-\infty}^{x} f(z) \, dz$$

is the cdf corresponding to $f(X)$. Since $F(X)$ is defined over the range 0 to 1, a particular value giving $F(X_0)$ is obtained with uniform probability by sampling from $U(0, 1)$. That is, $r_0 = F(X_0)$ and X_0 is uniquely determined by r_0, that is, $X_0 = F^{-1}(r_0)$ (see Figure 5-2).

$$r_0 = F(X_0) = \int_{-\infty}^{X_0} f(z) \, dz$$

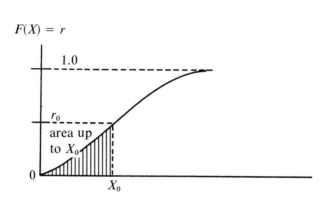

$F(X) = r$

Figure 5-2. Cumulative distribution function, cdf. X_0 is obtained by the inverse mapping of R_0. Solve for X_0 to obtain the inverse.

Any point within the area enclosed by a pdf can be selected at random with equal probability, that is, r_0. Then

$$\text{Prob}(X \le X_0) = F(X_0) = \text{area up to } X_0$$

$$F(X_0) = \text{Prob}[r_0 \le F(X_0)] = \text{Prob}[F^{-1}(r_0) \le X_0]$$

and consequently, $F^{-1}(r)$ is a variable that has $f(X)$ as its pdf. Solving for X_0 in terms of r_0 yields variates according to the pdf.

This result is sometimes called the *fundamental principle of Monte Carlo*. It is worthwhile to view the principle from several vantage points. Let's assume a discrete distribution with event probabilities of p_i, $i = 1, \ldots, n$. The distribution of sums is exactly $U(0, 1)$:

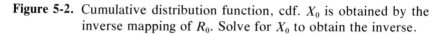

$$p_1 + p_2 + \ldots + p_{i-1} \le r < p_1 + p_2 + \ldots + p_i$$

where $r = $ uniformly random variate (RNG). That is, if E_i is likely to occur with probability p_i, then the random number r must fall between $\Sigma_{j=0}^{i-1} p_j$ and $\Sigma_{j=0}^{i} p_j$.

The same result is possible in the continuous case when the following theorem is used.

THEOREM. If $u = h(x)$ is a one-to-one transformation and $f(x)$ is a pdf (probability density function) of x, then the pdf of u, $g(u)$, is

$$g(u) = f(x) \left| \frac{dx}{du} \right|$$

This is the "distribution of functions of random variables" theorem familiar to all alert statisticians. In the case where $|dx/du|$ is multivalued, the theorem must be modified to include these values:

$$g(u) = \sum_i f(x_i) \left| \frac{dx_i}{du} \right|$$

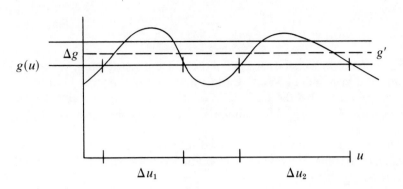

Figure 5-3. The multivalued distribution function $g(u)$.

where i sums over multivalues.

An informal proof is given, without loss of generality, from Figure 5-3.

$$\text{Prob}[g \leq g(u)] = \int_{-\infty}^{\infty} g(u) \, du = g(u') \, \Delta u$$

This, of course, is obtained by the mean value theorem.

The "typical" element, $g' \in \{g, g + \Delta g\}$, is an arbitrary value within the interval. Substitution of $\Sigma f(x_i) \, |dx_i/du|$ into the right-hand side yields

$$\text{Prob}[g \leq g(u)] = \sum_{i=1}^{2} f(x_i) \left| \frac{\Delta x_i}{\Delta u_i} \right|$$

and taking the limit as $\Delta u_i \to 0$ gives the final result:

$$g(u) = \sum_{i=1}^{2} f(x_i) \left| \frac{dx_i}{du_i} \right|$$

We wish to use this result to show that the distribution, $g(u)$, is uniform when $u = h(x)$ is a cdf of the pdf $f(x)$.

$$u = \int_{0}^{x(u)} f(s) \, ds$$

$$\frac{d}{du} \left[\int_{0}^{x(u)} f(s) \, ds \right] = \frac{d}{du} (u) = 1$$

$$f(x) \cdot \frac{dx}{du} = 1$$

$$\frac{dx}{du} = \frac{1}{f(x)}$$

Now, substitute $|dx/du|$ into the theorem and

$$g(u) = f(x) \cdot \left| \frac{dx}{du} \right| = f(x) \cdot \left| \frac{1}{f(x)} \right| = 1.$$

The pdf of $g(u)$ is exactly the pdf of the uniform distribution; hence, the fundamental principle of Monte Carlo simulation is shown to be true.

This result is also the basis of inversion sampling, as shown in the introduction. How does this technique compare by the binary criterion given earlier?

The *inverse transformation* is used to give good q generators when inversion is possible. However, an inversion has not been found for normal generators.

$$\text{random number} = F(X)$$

$$= \int_{-\infty}^{X} \frac{1}{\sqrt{2\pi}} e^{-t^2/2} dt$$

$$\text{Normal number} = F^{-1} \text{ (random number)}$$

Box and Muller [3] succeeded in almost inverting the normal cdf to obtain a closed expression for $X = F^{-1}$ (Rand) which gives good (quality = 7) normal random numbers.

Let U_1 and U_2 be random numbers on $(0, 1)$. Two normal deviates can be obtained through the transformation

$$X_1 = (-2 \log_e U_1)^{1/2} \cos (2\pi U_2)$$

$$X_2 = (-2 \log_e U)^{1/2} \sin (2\pi U_2)$$

The transformation of Box and Muller is based on the independence of X_1 and X_2. The joint distribution of two independent standarized normal deviates, i.e., both belonging to $N(0, 1)$, is

$$p(X_1, X_2) = \frac{1}{2\pi} \exp [(\tfrac{1}{2}X_1^2) + X_2^2)]$$

$$= \frac{1}{\sqrt{2\pi}} \exp (\tfrac{1}{2}X_1^2) \frac{1}{\sqrt{2\pi}} \exp (\tfrac{1}{2}X_2^2)$$

$$= p_1(X_1)p_2(X_2)$$

This independence of p_1 and p_2 motivates investigation into inverting p_1 and p_2 separately to obtain the desired deviates X_1 and X_2. Suppose

$$X_1 = r \cos \theta$$

$$X_2 = r \sin \theta$$

are used to transform the joint distribution $p(X_1, X_2)$ into $p(r, \theta)$. Then the probability density function becomes

$$p(r,\theta)\ dr\ d\theta \ = \ \frac{1}{\sqrt{2\pi}}\ \exp\left(\frac{-r^2}{2}\right)\ r\ dr\ d\theta$$

$$= \ \exp\left(\frac{-r^2}{2}\right)\ d(\tfrac{1}{2}r^2)\cdot\frac{1}{\sqrt{2\pi}}\ d\theta$$

$$= \ p_1(r)\cdot p_2(\theta)$$

Thus, r^2 and θ are independently distributed and can be sampled from $p_1(r)$ and $p_2(\theta)$ independently. The samples obtained say, r_0 and θ_0, can be used to find X_1 and X_2:

$$U_1 \ = \ \int_0^{r_0} p_1(z)\ dz = \int_0^{r_0} \exp\left(\frac{-z^2}{2}\right)\ d\left(\frac{z^2}{2}\right)$$

$$= \ -\exp\left(\frac{-z^2}{2}\right)\Big|_0^{r_0} \ = \ 1 - \exp\left(\frac{-r_0^2}{2}\right)$$

and since $1 - U_1$ and U_1 give like results,

$$r_0 = (-2\ \log_e\ U_1)^{1/2}$$

Second,

$$U_2 \ = \ \int_0^{\theta_0} \frac{1}{2\pi}\ dz = \frac{\theta_0}{2\pi}$$

$$\theta_0 \ = \ 2\pi\ U_2$$

So that

$$X_1 = r_0 \cos\ (2\pi\ U_2) = (-2\ \log_e\ U_1)^{1/2} \cos\ (2\pi\ U_2)$$

$$X_2 = r_0 \sin\ (2\pi\ U_2) = (-2\ \log_e\ U_1)^{1/2} \sin\ (2\pi\ U_2)$$

The form given provides a compact, straightforward method that will be valuable for most applications. In particular, this technique yields normal deviates accurate even in the tail regions of $N(0, 1)$.

5-4 Some Distributions

The partial inversion method of Box and Muller previously given is only one example of inversion sampling. The inversion method is useful only where inversion formulas are obtainable. Often approximate inversion is possible where exact inversion is impossible due to integration complexity; as an example, consider Kahn's approximation.

Normal Distribution

The Kahn approximation to the normal distribution is obtained by an approximation formula. Let

$$e^{-x^2/2} \doteq \frac{2e^{-kx}}{(1 + e^{-kx})^2} \qquad k = \sqrt{\frac{8}{\pi}}$$

be an approximation to the exponential. The normal distribution is obtained by inversion:

$$\text{RNG} = \int_0^x \frac{2ke^{-kx}}{(1 + e^{-kx})^2} \, dx = \frac{2}{1 + e^{-kx}} - 1$$

Thus the random variate from $N(0, 1)$ is X:

$$X = \frac{1}{k} \log_e \frac{1 + \text{RNG}}{1 - \text{RNG}}$$

Hence, to sample from $N(0, 1)$, simply compute X. This method is easier to program than the method of Box and Muller because only one formula is calculated. The technique is faster because only a logarithm (not a logarithm, sin/cos, and square root) is needed. The method of Box and Muller would be given by an algorithm as follows:

$$X \leftarrow \text{IF "even" THEN } X_1 \text{ ELSE } X_2$$

This means that the random variate X is alternately X_1 or X_2 each time the program is called.

The Kahn formula is an approximation which gets worse as the tails are lengthened [2]. The accuracy decreases in the tails, and therefore this method should not be used for experiments needing high accuracy. Other techniques for sampling from $N(0, 1)$ will be given later.

Exponential Distribution

The exponential distribution is integrable and so can be obtained by inversion:

$$p(x) = \frac{1}{\sigma} e^{-x/\sigma}$$

$$\text{RNG} = \int_0^x \frac{1}{\sigma} e^{-x/\sigma}$$

$$X = -\sigma \cdot \log_e (\text{RNG})$$

where σ = average and RNG = 1 $-$ RNG statistically.

A computer program to compute an *exponential variate X* is simply a one-line assignment statement in most high-level programming languages (providing that the logarithm routine is built in).

The exponential function is useful for computing random variates from other distributions too. The following sampling algorithms are based on exponential variates.

Gamma Generator

The gamma generator is obtained by sampling exponential variates and then computing their sum. Let e_i be exponential variates with mean of μ.

$$G = \sum_{i=1}^{k} e_i$$

$$G = \text{variate from gamma}$$

Since the e_i are computed from a logarithm,

$$G = -\sum_{i=1}^{k} \log_e (RNG_i) * \mu$$

$$= -\mu \log_e \prod_{i=1}^{k} (RNG_i)$$

due to the equivalance of "sum of logs equal log of products."

The mean value M_G and variance V_G are obtained from the gamma formula:

$$F(G) \sim \frac{(1/\mu)^k G^{k-1} e^{-G/\mu}}{(k-1)!}$$

$$M_G = k\mu$$

$$V_G = k * \mu^2$$

Therefore, the gamma generator is one which multiplies k random numbers from $U(0, 1)$ and performs the $\log_e$.

Chi-square Generator

The chi-square is related to the gamma, so that it, too, is obtained from previous results. Set $\mu = 1/2$ in the distribution. Then

$$k = \frac{\text{mean}}{2}$$

Thus, the X^2 generator becomes

$$X^2 = \begin{cases} \text{Gamma, } \mu = \frac{1}{2} & \text{if } k = \text{integer} \\ \text{Gamma, } \mu = \frac{1}{2}, \text{ plus } z^2 & \text{if } k \neq \text{integer} \end{cases}$$

where $z \sim N(0, 1)$ variate.

Thus, the normal, gamma, and uniform generators are needed to compute a X^2 variate.

Beta Generator

The beta generator uses two gamma variates as follows:

$$\beta = X_1/X_1 + X_2 \qquad 0 \leq \beta < 1$$

The (X_1, X_2) pair is obtained from gamma generators with distinct (k_1, k_2).

$$k = k_1 + k_2$$

The mean μ must be the same for both gamma generators.

Cosine Generator

The cosine-generator is directly obtained by using the theorem in the previous section. The technique is given here, solely to show how "homemade" generators are constructed.

Suppose $X = \cos \theta, 0 \leq \theta \leq 1$. The value of θ is selected randomly, and so we are essentially computing a random cosine, $X = \cos (\text{RNG})$ and $g(x) = f(\theta) |d\theta/dx|$ from the theorem. The density of θ is 1 because it is uniformly distributed. The inverse of X is $G = \cos^{-1} X$, and

$$\left| \frac{d\theta}{dx} \right| = \frac{1}{\sqrt{1 - x^2}}$$

We will show later where it may be better to sample from $1/\sqrt{1-x^2}$ than from $\cos \theta$.

Spheres

The normal density function has constant probability on the surfaces of n-dimensional spheres with common centers [4]. This fact can be used to generate points uniformly on the surface of spheres in n dimensions. The technique points out how clever use of known sampling generators can lead to simple generators for rather complex problems.

The sphere generator is carried out by the following procedure:

(1) Generate n independent normal deviates, X_i, $i = 1, 2, \ldots, n$.

(2) Locate the point y on the unit n-sphere by computing direction cosines corresponding to the point $X = (X_1, X_2, \ldots, X_n)$:

$$\frac{X_i}{\left(\sum_{i=1}^{n} X_i^2 \right)^{1/2}} \quad : \quad i = 1, 2, \ldots, n$$

The points y obtained from this generator will be uniformly distributed over the sphere.

EXAMPLE Let $n = 2$ in the sphere generator just given. The normal deviates (X_1, X_2) yield one point on the circumference of a circle. See Figure 5-4.

5-5 Discrete Sampling

The inversion technique is easily applied to discrete pdf's when closed-form summations are possible. When they are not possible, it is still sometimes practical to look up the sums of probabilities in tables.

EXAMPLE See Figure 5-5. A table of values is stored for the cdf shown in the figure. The RNG number that corresponds with 0, 1, 2, 3, or 4 is then looked up.

X	RNG number
0	0.06
1	0.31
2	0.69
3	0.94
4	1.0

For example, RNG = 0.55 corresponds with $X = 2$ because it is between 0.31 and 0.69.

In another sense, it is easy to see that $X = 0$ will be sampled 6 percent of the time, $X = 1$ 25 percent, $X = 2$ 38 percent, $X = 3$ 25 percent, and $X = 4$ 6 percent of the time. This method can be extended to the method of composition which will be given later.

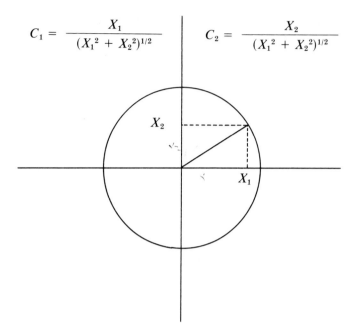

$$C_1 = \frac{X_1}{(X_1{}^2 + X_2{}^2)^{1/2}} \qquad C_2 = \frac{X_2}{(X_1{}^2 + X_2{}^2)^{1/2}}$$

Figure 5-4. Location of random point (X_1, X_2) on the surface of a 2-sphere.

Poisson Distribution

n is Poisson-distributed if we notice that n = the number of events occurring in unit of time t, where t is distributed as an exponential variate. Therefore, to generate $t_i = -\log r_i$, $r_i \in U(0, 1)$, compute

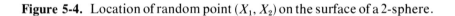

$$\sum_{i=1}^{n} t_i \leq \lambda \leq \sum_{i=1}^{n+1} t_i$$

where

$$p_n(\lambda) = \frac{\lambda^n}{n!} e^{-\lambda} \qquad \text{(Poisson distribution)}$$

Since $t_i = -\log r_i$, we substitute into the generator

$$-\sum_{i=1}^{n} \log r_i \leq \lambda \leq -\sum_{i=1}^{n+1} \log r_i$$

$$-\log \prod_{i=1}^{n} r_i \leq \lambda \leq -\log \prod_{i=1}^{n+1} r_i$$

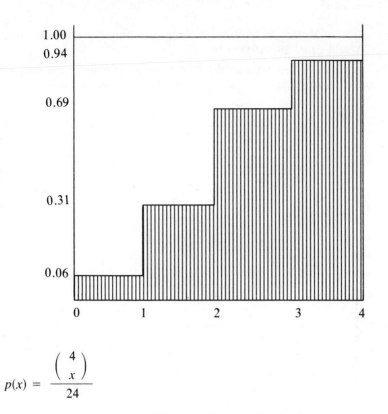

$$p(x) = \frac{\dbinom{4}{x}}{24}$$

Figure 5-5. Cdf of a discrete density function

or

$$\prod_{i=1}^{n} r_i \le e^{-\lambda} \le \prod_{i=1}^{n+1} r_i$$

The algorithm is to compute products $r_1, r_2, r_3, \ldots, r_n$, until it exceeds $e^{-\lambda} =$ constant. The expected value of n is λ, and the efficiency is $1/\lambda$.

References

[1] Teichroew, D. Distribution sampling with high speed computers. Ph.D. thesis, North Carolina State College, 1953.
[2] Tocher, K.D. *The Art of Simulation*. The English University Press, London, 1963.

[3] Box, G. and Muller, M. A note on the generation of normal deviates. *Ann. Math. Stat.* 28: (1958) 610.

[4] Muller, M.E. A note on a method for generating points uniformly on *N*-dimensional spheres. *Comm. ACM.*, 2, 4, (1959), p. 14-20.

6   **Rejection Sampling**

6-1   The Method

It is frequently simple to calculate the pdf from which sampling is to be done, but difficult to evaluate the cdf or its inverse. Instead, a "rejection" technique is possible whereby the probability that a uniform variate X does not exceed y is precisely y:

$$y = \frac{pdf}{M} \qquad 0 \le pdf = f(x) \le M$$

The scaling factor M is selected to normalize the pdf, so that $0 \le y \le 1$ is a probability. A test value R is chosen from $U(0, 1)$, and X is rejected if $R > y = f(x)/M$; the distribution of accepted x's will be $f(x)$. This follows since

$$\text{Prob}(x \text{ accepted} \mid x, R) = \text{Prob}(R \le y) = y \, dx$$

and

$$\int_0^1 y \, dx = \frac{1}{M} \int_0^1 f(x) \, dx = \frac{1}{M}$$

since $f(x)$ is a pdf. The number of trials before a successful pair is found (x accepted) has a geometric distribution:

$$p(n) = E(1 - E)^{n-1} \qquad E = 1/M$$

The mean value of $p(n)$ is $1/E = M$. Therefore, M is a measure of rejection efficiency. M trials are expected before an x is accepted.

To see clearly why the rejection technique works, study the outline of the proof given below. An arbitrary density function $f(x)$ is shown.

The reader should be aware of the limitations of the method. First, the pdf, $f(x)$, must be bounded. The value of M must be finite. Second, the efficiency of the method equals the proportion of the area that falls under $f(x)$. If $f(x)$ does not fill up the square (region, in general), then the method is wasteful of RNGs. See Figure 6-1.

Let x_0, y_0 be uniform random numbers on $(0, 1)$.

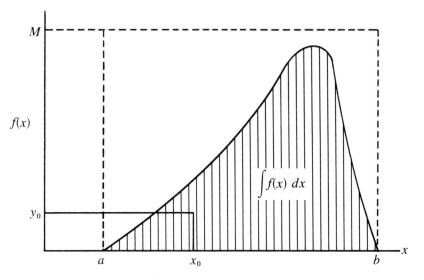

Figure 6-1. An arbitrary pdf contained within a region

$$\text{Prob } (x_0, y_0 \text{ accepted}) = \frac{\displaystyle\int_a^b f(x)\, dx}{M(b-a)}$$

$$= \frac{1}{M(b-a)};$$

$$\text{since } \int_a^b f(x)\, dx = 1$$

$$\text{Prob } (x_0 \text{ acceptable}) = \frac{\text{Prob } (x_0 \text{ and acceptance})}{\text{Prob (acceptance)}}$$

$$= \frac{\dfrac{1}{(b-a)} \cdot \dfrac{f(x_0)}{M}}{\dfrac{1}{M(b-a)}}$$

$$= f(x_0)$$

In summary, the algorithm for rejection sampling is as follows:

(1) Find a constant, M, such that $Mf(x) \leq 1$, $a \leq x \leq b$.

(2) Let x be obtained from a uniformly random r_1:

$$x = a + (b-a)r_1$$

(3) Accept x if it is true that [with r_2 also from $(0, 1)$]

$$r_2 \leq M f [a + (b - a)r_1] = M f(x)$$

(4) If x in (3) is accepted, stop; otherwise, sample a new r_1, r_2 and repeat (3).

Suppose we want to produce random numbers z in $(-1, 1)$ according to

$$f(x) = \frac{1}{\pi(1 - x^2)^{1/2}} \qquad -1 < x < 1$$

using the cdf to obtain an inverse:

$$R = \int_0^z \frac{dx}{\pi(1 - x^2)^{1/2}} = \frac{\arcsin z}{\pi}$$

Then

$$z = \sin(\pi R) \qquad R \text{ is from } U(0, 1)$$

The calculation of sine can be avoided, however, by using rejection instead.

Select two uniform variates X and Y from $U(0, 1)$. X and Y define a point within a unit square as shown in Figure 6-2. To make sure that $\theta = \arctan(X/Y)$ is uniformly distributed on $(0, \pi/2)$, reject the pair if $X^2 + Y^2 > 1$. That is, let $f(x)/M = \sqrt{(1 - X^2)}$, and the test becomes $Y > \sqrt{(1 - X^2)}$:

$$Y^2 > 1 - X^2$$

$$Y^2 + X^2 > 1$$

If $X^2 + Y^2 \leq 1$, the pair is accepted and $\sin \theta = X/\sqrt{X^2 + Y^2}$. The square root may be avoided by noting that

$$\sin(2\theta - \pi/2) = -\cos^2 2\theta = \frac{X^2 - Y^2}{X^2 + Y^2}$$

which has the same distribution. The efficiency is clearly $\pi/4$.

Later, the rejection procedure will be combined with other methods to expedite steps in producing normal deviates.

EXAMPLE Determine the sampling routine for

$$x = \sqrt{RNG}$$

that is, the square root generator. By noting

$$X^2 = RNG = \int_0^X 2s \, ds$$

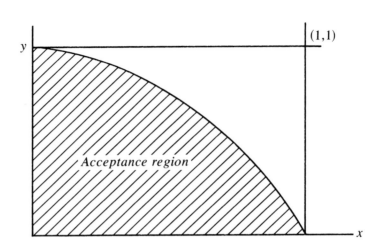

Figure 6-2. Acceptance area $= \pi/4$

the plot shown in Figure 6-3 is realized.

Let $p(x) = 2x/M^2, 0 \le x \le M$, and then select random coordinates from the interval $(0, M)$. Since pairs of random numbers locate a point which either falls in the shaded area or does not, the rejection algorithm is exactly a maximum function:

$$X = \max \{r_0, r_1\}$$

$$r_0 = M \cdot RNG$$

$$r_1 = M \cdot RNG$$

Power Function

The square root generator can be generalized to the power generator. Hence, to sample from

$$f(x) = (n + 1)X^n \qquad 0 \le X < 1$$

we must generate random numbers

$$RNG_1, RNG_2, \ldots, RNG_{n+1}$$

and select the largest RNG_i for the value X.

$$X = \max_{i \le n+1} \{RNG_i\}$$

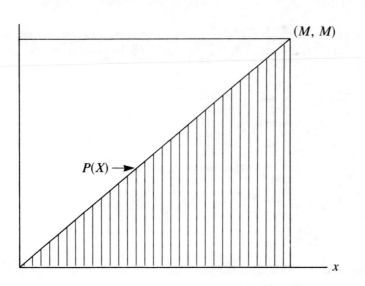

Figure 6-3. Density of the square root generator

The efficiency of the power generator is $1/(n + 1)$. To compute the negative powers of a power generator, merely let X be the reciprocal of the value above.

$$X = \frac{1}{\max \{RNG_i\}}$$

6-2 Some Generators

Normal Batchelor Technique

The Batchelor mixed rejection technique is sometimes more appropriate than the straightforward method of Box and Muller (fast, less portable). The technique is given now and a FORTRAN subroutine is given in Figure 6-4. Let

$$
\begin{aligned}
p(x) &= \frac{1}{\sqrt{2\pi}} \exp\left(\frac{-x^2}{2}\right) \\
&= \frac{1}{\sqrt{2\pi}} \left\{ \exp\left(\frac{-x^2}{2}\right) r(x) \right. \\
&\quad \left. + \tfrac{1}{2} \exp\left[-\frac{(x - 2)^2}{2} \right] S(x) \right\}
\end{aligned}
$$

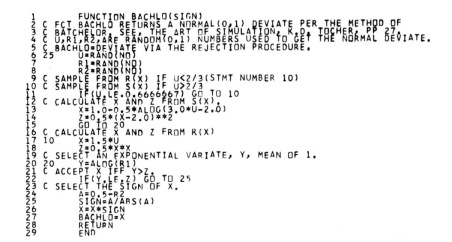

Figure 6-4. FORTRAN subprogram for Batchelor normal deviate generator

where

$$r(x) \;=\; \begin{cases} 1 & 0 \le x \le 1 \\[4pt] 0 & x > 1 \end{cases}$$

and

$$S(x) \;=\; \begin{cases} 0 & 0 \le x - 1 \\[4pt] 2 \exp\left[-2(x-1)\right] & x > 1 \end{cases}$$

$$\int_{-1}^{1} p(x)\,dx = 0.6826$$

Thus, two parts compose $p(x)$: first, an exponential part with $-x^2/2$ as argument over the range $(0, 1)$; second, the exponential part $-(x-2)^2/2$ over the remainder. Since

68 percent of the time $r(x)$ is used and 32 percent of the time $S(x)$ is used. Sampling from both is performed by rejection.

ALGORITHM

(1) Choose random numbers U, V, W from $(0, 1)$ and do either (2a) or (2b).

(2) (a) If $U \le 0.6826$, let $x = U/0.6826$ and $z = x^2/2$ otherwise, [this is set up for rejection to obtain $\exp(-x^2/2)$].

   (b) If $U > 0.6826$, let $x = 1 - \frac{1}{2} \log_e (3U - 2)$ and $z = \frac{1}{2}(x - 2)^2$. (This sets up rejection for $\exp -(x - 2)^2/2$.

(3) Select Y from V: $Y = - \log_e V$

(4) Accept x if $Y > z$; otherwise return to (1) and try again. (Acceptance rate = 78 percent.)

(5) Use W to select a sign: plus if $W > \frac{1}{2}$, minus if $W < \frac{1}{2}$.

Improvements to the Box and Muller Method

The Box and Muller method can be improved in speed by eliminating the lengthy square root and trigonometric calculations.

Bell [1] has suggested replacing the sin/cos calculations with a von Neumann rejection [2] technique. The "rejection" technique avoids direct calculation of cdf's and sometimes pdf's in sampling from a pdf.

The steps in the rejection procedure for $\cos (2\pi R)$ are as follows:

(1) Select two uniform variates R_0 and R_1.

(2) Calculate $\quad \cos (2\pi R) + \cos^2 (2\pi R) - \sin^2 (2\pi R) = \dfrac{R_0^2 - R_1^2}{R_0^2 + R_1^2}$

(3) Reject $\cos(2\pi R)$ unless $R_0^2 + R_1^2 \le 1$.

The efficiency of this procedure is $\pi/4$, that is, $100(1 - \pi/4)$ percent of the numbers R_0 and R_1 generated are rejected, while $100(\pi/4)$ percent are accepted. Cosine is obtained in this manner by inverting the pdf, $F(x) = 1/(\pi\sqrt{1 - X^2})$, $0 \le X \le 1$:

$$F(X) = \int_0^X \frac{dx}{\pi\sqrt{1 - z^2}} = R, \quad \text{where } R \text{ is uniform on } (0, 1)$$

$R = (\cos^{-1} x)/\pi$; therefore, $X = \cos (\pi R)$. X is obtained by rejection from $F(X)$, thus avoiding calculation of cosine.

6-3 Variance Reduction

Monte Carlo integration is closely related to rejection because it deals with areas, as does rejection. In order to reduce error and efficiently utilize RNGs, the Monte Carlo technique is sometimes modified. Let J be the value of an integral to be calculated:

$$J = \int_a^b f(x)\ dx \quad \text{(true value)}$$

L/N = estimated value, where L = number of points within the area being integrated and N = total samples

A variation of the central limit theorem gives a bound for the error in integration by Monte Carlo:

$$\left| \frac{L}{N} - J \right| \le X_p \sqrt{\frac{\text{var}\ (J)}{N}}$$

where X_p is the critical value obtained from the student t distribution.

$$\text{var}\ (J) = (b - a) \int_a^b f^2(x)\ dx - J^2$$

Now, since $f(x)$ is bounded by one side of an enclosing rectangle of height c,

$$\int_a^b f^2(x)\ dx \le cJ$$

Therefore, the bound on variance is

$$\text{var}\ (J) \le c(b - a)J - J^2$$

The error in $|L/N - J|$ is reduced by reducing var (J). This is done in several ways. One obvious method is to reduce the encompassing rectangle, of area $c(b - a)$.

Regular Parts

The reduction in variance by regular parts consists of sectioning the encompassing rectangle, as shown Figure 6-5. The rectangle is made smaller, and $c(b - a)$ is reduced. The variance is reduced, and greater accuracy results.

Group Sampling

Group sampling is a simple method for variance reduction. The idea is to take more samples in the region of rapid change [derivative of $f(x)$].

Figure 6-6 shows a typical group sampling section. The samples are taken in groups. The greatest number of samples should be taken from

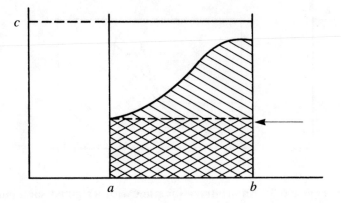

Figure 6-5. An arbitrary function with a regular part

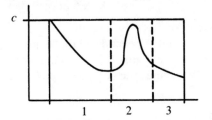

Figure 6-6. An arbitrary function with three groups.

group 2. The rectangles are not reduced, but the samples are used where needed. Therefore, more information is gained about the rapidly changing function.

Importance Sampling

Computing the area of a function $f(x)$ by Monte Carlo simulation can be improved by weighting the samples. The uniform distribution gives no advantage to "important" samples. If the distribution

$$p(x) = \text{pdf of samples}$$

is used, then the integration is in error because it violates our assumption of uniformity. This, however, is overcome by changing the test criterion.

$$J = \int_a^b f(x)\, dx = \int_a^b \frac{f(x)}{p(x)}\, p(x)\, dx$$

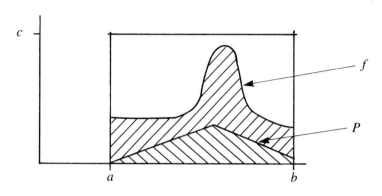

Figure 6-7. An arbitrary function with weighted sampling.

$$\text{Area} \doteq \frac{1}{N} \sum_{i=1}^{N} \frac{f(x_i)}{p(x_i)}$$

Test $f(x_i)/p(x_i)$ instead of $f(x_i)$.

The selection of $p(x)$ must be made so that a greater number of samples fall where most needed. See Figure 6-7. The sampler pdf is contoured to the function $f(x)$. In the extreme, the sampler equals the function:

$$p(x) = f(x)$$

When this happens, the advantages are lost because the sampling consumes more time and effort than the accuracy it provides. Therefore, $p(x)$ should be a simple distribution like the triangular pdf. The triangular pdf is sampled by computing the sum or difference of two RNGs.

References

[1] Bell, J.R. Algorithm 334—Normal random deviates. *Comm. ACM,* 11(7): 498 (July 1968).
[2] von Neumann, J. Various techniques used in connection with random digits. *Natl. Bur. Std. Appl. Math.,* 36, Sept. 12, 1959.

7

Other Techniques

7-1 Introduction

The q generators given up to this point are commonly used in simulation, in Monte Carlo calculations, and in scientific applications requiring "noise." All these employ techniques which may be applied to any distribution. Special techniques may be appropriate under severe time or memory restrictions, however, and other methods are sought.

The composition and parallel processor techniques given here apply to speed-restricted or portability-restricted generators.

7-2 General Composition

Composition techniques sample from a pdf by composing many density functions which approximate the desired pdf:

$$pdf = f(X) = \sum_n g_n(X)P_n$$

The functions g_n are selected with probability P_n and constitute a "best" fit to $f(X)$. One or more of the functions g_n may be step function(s) (i.e., a table of values) which partially describe $f(X)$ or which describe $f(X)$ within a specified range of values. For machine calculation, composition methods are usually quite fast but require sizable memory.

Several procedures based on the composition technique were developed by Marsaglia and others [1, 2]. The storage requirement and complexity of these procedures can be justified only where speed is of prime importance (rated 10).

For example, the density function $f(X) = (1/\sqrt{2\pi}) \exp(-X^2/2)$ is approximated over selected intervals of the x axis so that simpler functions can be most often used. Then a central limit approach can be used to find X during 86.38 percent of the time if

$$X = 2(U_1 + U_2 + U_3 - 1.5)$$

where $\sqrt{K/12} = \sqrt{1/4} = 1/2$, since $K = 3$ and accuracy to within $\sqrt{3K} = 3$ units from the mean is desired [$f(3) = 0.86$]. Then 13.62 percent of the time X must be sampled from $f(X) - 0.86638g_1(X)$ which can be closely fitted by

Figure 7-1. $g_1(X)$: a table of steps which approximate the normal $= X + 0.1U$, where U is uniform on $(0, 1)$ and $\hat{X}$ is discrete on X.

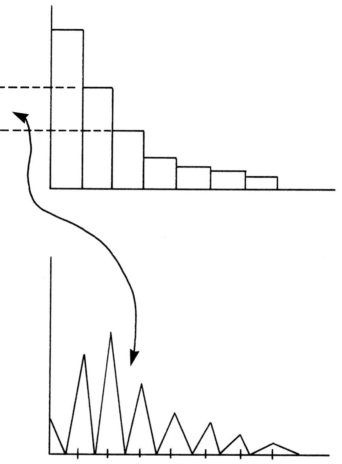

Figure 7-2. $g_2(X)$: a set of triangular-distribution approximations to the difference between $g_1(X)$ and $p(X)$. The desired "tooth" is selected from sampling g_1, and then X is in turn obtained from sampling a "tooth" of $g_2(X)$.

a triangular density $g_2(X)$; 11.07 percent of the time $X = 1.5(U_1 + U_2 - 1)$. This leaves 2.55 percent to be sampled from $g_3(X)$, and so forth.

It is clear that $1.5(U_1 + U_2 - 1)$ is distributed as a triangle density, however, to obtain X; 2 percent of the time much more work is involved in fitting the tail region. A combination of the Box and Muller method and the rejection $(X_1^2 + X_2^2 < 3)$ method is used.

Probably the fastest normal generator uses three functions as follows [2] (see also Figures 7-1 to 7-3).

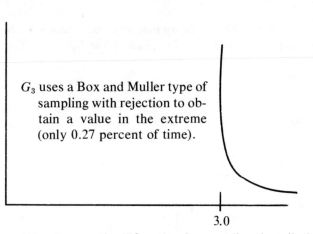

G_3 uses a Box and Muller type of sampling with rejection to obtain a value in the extreme (only 0.27 percent of time).

3.0

Figure 7-3. $g_3(X)$: a "correction" function for sampling the tail of $N(0,1)$.

A sum of approximations to $p(t) = (1/\sqrt{2\pi}) \exp(-t^2/2)$ is found as

$$p(X) \doteq 0.9578g_1(X) + 0.0395g_2(X) + 0.0027g_3(X)$$

so that 95.78 percent of the time $g_1(X)$ is used, 3.95 percent of the time g_2 is used, and 0.27 percent of the time g_3 is used. First, $g_1(X)$ is used to obtain X within a range, say Δ. Then greater refinement is obtained by sampling from $g_2(X)$. For values greater than 3.0, sampling is done entirely by $g_3(X)$.

The step function $g_1(X)$ is actually a table containing a few hundred entries (Figure 7-1). Looking up entries in the table lends itself best to machine language coding and contributes to the complexity of such a method, but it should be remembered that speed is the main concern when this technique is used.

The function $g_2(X)$ provides a correction to the table values first by sampling for one of the "teeth" and then by locating the value of X along its slanted side (Figure 7-2).

As always, the tail region offers the greatest difficulty in sampling. The analytic approach of Box and Muller is modified with a rejection procedure to provide a highly accurate but slow method. If $r = (9 + 2W)/(U_1^2 + U_2^2)$, then two normal deviates are obtained from the tails (Figure 7-3):

$$X_1 = U_1r \quad X_2 = U_2r$$

7-3 Central Limit Approach

A method unique to the normal distribution and, therefore, not applicable to other distributions is the central limit approach. It is a fair generator (rated 5) and is mentioned here for completeness, as well as for its potential for parallel computers.

Suppose that $r_1, r_2, \ldots, r_k$ are independent random numbers on $(0, 1)$, each with expectation $E(r_i) = U$, variance var $(r_i) = \sigma^2$. Then

$$\lim_{k \to \infty} \text{Prob}(a < Z < b) = \frac{1}{\sqrt{2\pi}} \int_a^b \exp\left(\frac{-z^2}{2}\right) dz$$

where

$$Z = \frac{\sum_1^k r_i - ku}{\sigma\sqrt{k}} \qquad Z \text{ is a normal deviate}$$

$$E\left(\sum_1^k r_i\right) = ku$$

$$\text{var}\left(\sum_i^k r_i\right) = k\sigma^2$$

7-4 Sampling from the Normal Distribution in Parallel Processor Machines

Normal deviates, or numbers sampled from $p(x) = (1/\sqrt{2\pi})e^{-x^2/2}$ are distributed with mean of zero and standard deviation equal to 1, $N(0, 1)$. The x's can be used in many areas of simulation and nonparametric statistics. Monte Carlo calculation of normal deviates can be efficiently and quickly obtained in parallel processor machines like ILLIAC IV, by utilizing pseudorandom uniform number generators in each of many processor elements (abbreviated PE) [3, 4].

Method

An appropriate method must involve parallel calculations throughout the process in order to take full advantage of multiple PEs. Several algorithms suitable for serial machines exist, but they do not allow overlapping or duplication during generation. For instance, the Box and Muller formulas use two uniform random numbers R_1 and R_2 to obtain two normal deviates N_1 and N_2.

$$N_1 = \sqrt{-2 \ln(1-R_1)} \cos (2\pi R_2)$$
$$N_2 = \sqrt{-2 \ln(1-R_1)} \sin (2\pi R_2) \tag{7.1}$$

It would be difficult, if not impossible, to use this popular generator and still realize the full power of a parallel computer.

Another generator which is serially inferior to other normal deviate generators uses the central limit theorem in providing deviates within $\sqrt{3k}$ units of the mean.

$$N = \frac{\sum_{i=1}^{k} x_i - \dfrac{k}{2}}{\sqrt{k/12}} \qquad (7.2)$$

where N = normal deviate, distributed on $N(0, 1)$

x_i = ith pseudorandom number distributed on $U(0, 1)$

k = the number of samples

The possibilities for simultaneous computation are immediately obvious in this formulation. In Equation (7.2), the k random numbers may be simultaneously generated and then routed to a single PE as the sum in Equation (7.2). A subtraction of $k/2$ and division by $\sqrt{k/12}$ provide the final result.

Further Considerations

This method can be explored further to gain computational efficiency if the value of k is selected judiciously. In addition, the method used to obtain the X_i's must be chosen carefully to guarantee suitable serial correlation.

If $\sqrt{k/12}$ is made up of a power of 2, then division can be replaced with a much faster shift.

$$\sqrt{k/12} = 2^l \qquad \left\{ \begin{array}{l} l = 0,\ k = 12 \\ l = 1,\ k = 48 \\ l = 2,\ k = 192 \\ l = 3,\ k = 768 \end{array} \right.$$

Setting k to 48 yields

$$N = \tfrac{1}{2} \sum_{i=1}^{48} x_i - 12 \qquad (7.3)$$

However, only 48 PEs are used, which leaves 25 percent of the machine idle. Thus, if k is 64, the denominator becomes 1.909756 and

$$N = \frac{\sum_{i=1}^{64} x_i - 32}{1.909756}$$

which allows deviates within 13.9σ of the mean.

The trade-off between division shift and machine utilization must be a consideration in the generator implementation.

To obtain the random numbers x_i, a recursion is desired, so that subsequent numbers may be obtained from some previous number. The recursion is made in steps of 48 or 64 since this is the number of numbers to be generated each time the algorithm is applied. This means that the new number is advanced 48 or 64 terms ahead in the sequence for each PE. The Lehmer congruence generator is ideal since it allows the advance feature sought.

$$x_i \equiv ax_{i-1} \quad (\text{mod } m) \quad i = 1, 2, \ldots \quad (7.4)$$

Proper selection of m and a ensures sufficient cycle length of the generator (see Part 1).

$$a = p$$

where p is relatively prime to $m-1$ and relatively prime to $2^m - 1 =$ cycle length

m is a prime number.

The selection of m is made from Mersenne numbers, depending upon the word size of a particular machine. However, this form of the recursion does not advance 48 terms in the sequence as prescribed.

$$X_1 \rightarrow X_{49} \rightarrow X_{97} \rightarrow \ldots$$

Repeated applications of Equation (7.4) will ultimately yield the desired sequence:

$$X_{49} \leftarrow aX_{48} \leftarrow a^2X_{47} \leftarrow \ldots \leftarrow a^{48}X_1$$

Thus, the advance feature is realized if a new constant multiplier is used

$$X_{n+48} \equiv a^{48}X_{n-1} \quad (\text{mod } m) \quad n = 1, 2, \ldots \quad (7.5)$$

and it becomes

$$X_j \equiv AX_{j-1} \quad (\text{mod } m) \quad j = 1, 2, \ldots \quad (7.6)$$

$$A \equiv a^{48} \quad (\text{mod } m) \quad \text{or } a^{64} \quad (\text{using the 64 option})$$

Equation (7.6) is applied in each of 48 or 64 PEs, once for every normal deviate generated.

The starting value for each PE can be obtained from one arbitrary starting value and the 47 subsequent numbers obtained from Equation (7.4).

Steps in the Algorithm

(1) Generate 48 X's (or 64) from

$$X_j \equiv AX_{j-1} \quad (\text{mod } m)$$

simultaneously in PEs.

(2) Shift each number in each PE 1 bit to the right (skip if using the 64 option).

(3) Normalize and float (in each PE).

(4) Bring together all numbers in a sum.

(5) Subtract $k/2$ from sum (and divide if using the 64 option).

Implementation on ILLIAC IV

ILLIAC IV provides hardware for a variety of configurations in parallel processing and word size. The 64 PEs in each quadrant are more than sufficient for the generation of 48 random numbers. Each PE can operate with 64-bit or twin 32-bit floating-point registers.

The value of m depends upon selection of the 64- or 32-bit configuration.

If 64 is used, $2^{61} - 1$ is the nearest Mersenne number available; however, if 32-bit hardware is used, $2^{31} - 1$ is ideal, allowing for the sign bit. Therefore, the 32-bit configuration is selected, which gives a cycle of $(2^{31} - 1)/48 > 2^{25}$ numbers.

The 64 PEs of one quadrant of the ILLIAC IV are used. If only 48 are used, the last 16 are set to zero. By using the procedure *random*, which is the coded Lehmer generator, floating-point pseudorandom numbers on the interval $(0, 1)$ are generated simultaneously across the PEs, which satisfies steps (1) to (3) of the algorithm. The X's are then summed by using a sum routine [5]. The number of additions required is 6, compared with 47 additions in a serial machine.

The TRANQUIL (high-level programming language) for one random normal deviate is as follows:

procedure NRMDEV (dev, random);

REAL dev; REAL PROCEDURE random; REAL ARRAY x {1:64} INTEGER i, j, k;

COMMENT the procedure random will supply a random element from a large population of real numbers uniformly distributed over the interval $0 < r < 1$;

begin

    FOR (i) SIM (1, 2, ..., 64) DO a {i}: = 0.0;

```
        FOR (i) SIM (1, 2, ..., 48) DO a {i}: = random;
        FOR (k) SEQ (0, 1, 2, 3, 4, 5) DO BEGIN j : = 2 ↑ k
        FOR (i) SIM (1, 2, ..., 64) DO a {i}: = a {i} + a {(i + j) mod 64} END;
        dev : = a {1} − 12.0
    END
    END NRMDEV
```

References

[1] Marsaglia, G., and Bray, T.A. A convenient method for generating normal variables. *Comm. ACM*, 6 (3): 260-264 (July 1964).

[2] Marsaglia, G., MacLaren, M.P., and Bray, T.A. A fast procedure for generating normal random variables. *Comm. ACM*, 7 (1): 4-10 (Jan. 1964).

[3] Box, G. and Muller, M. A note on the generation of normal deviates. *Ann. Math. Stat.*, 610, (1958).

[4] Winje, G.L. Random number generators for ILLIAC IV. Dept. of Computer Science, University of Illinois at Urbana. June 1969.

[5] Kuck, D.J. ILLIAC IV software and applications programming. *IEEE Trans. on Computers*, C17(8): 758-770 (Aug. 1968).

8 Conditional Bit Sampling

8-1 Introduction

The probability-integral transformation (inversion method) assures that if X is sampled from cumulative probability distribution (cdf) G, then $U = G(X)$ is uniformly distributed. Thus, if cdf G is to be sampled (on a computer), a random (pseudorandom) number is selected (computed) and $G^{-1}(U) = X$ is evaluated. The value X is sampled from G, or when a pseudorandom number, U, is used, X is said to have a pseudo-G distribution.

Computations of $G^{-1}(X)$ present a problem. Often direct analytic inversion of G is impossible or impractical. Several alternatives to avoid direct computation have been developed. Numerical inverse interpolation of exact G, numerical inverse interpolation in an approximation to G, or numerical inverse approximation to G^{-1} have been used. Generalized rejection techniques can be used to sample from G without computation of G^{-1} or without even performing the integration on the density of G. The method of mixtures (or composition) splits the probability density of G into a sum of simple probability densities. A method of subtrafuge can be used to create a process distributed similar to G.

Consider sampling from the normal distribution with expected value μ and standard deviation σ. Define

$$\phi(X) = \frac{1}{\sqrt{2\pi}} \int_{-\infty}^{X} e^{\frac{(t - \mu)^2}{2\sigma^2}} \tag{8.1}$$

and define $\phi(X)$ to be the same integral except that the lower limit is equal to 0 rather than $-\infty$. For all the sampling methods discussed in this book, $\mu = 0$ and $\sigma = 1$, since if X is sampled from the normal $\mu = 0$ and $\sigma = 1$, then $BX + A$ is from a normal with $\sigma = B$ and $\mu = A$.

Nearly all the above methods have been used to produce normal numbers [1]. For (almost) direct analytic inversion of the normal distribution, Box and Muller developed a method based on the classical technique for integration of $\phi(X)$ between $-\infty$ and ∞ to obtain unity [2, 3, 4]. The disadvantage of this method is that two pseudorandom numbers, a sine, a cosine, a natural logarithm, and a square root must be computed.

This chapter is taken from Rice (ed.), "Continuous Distribution Sampling: Accuracy and Speed," W. H. Payne and T. G. Lewis, Chapter 5.19 of Mathematical Software, New York: Academic Press, 1971.

A Pade-type, or rational, approximation to transform a uniform deviate to a normal deviate is not popular because faster procedures requiring fewer memory locations are available [5].

The central limit approach approximates a normal number by summing N uniform random numbers (subtrafuge method). It requires little memory space, is moderately fast and portable, but produces numbers restricted to $\pm\sqrt{3N}$ standard deviations of the mean. Teichroew approximated this distribution ($N = 12$) by curve fitting (Chebyshev polynomials), but the normal numbers are restricted to $\pm 4\sigma$ [1, 6].

Several composition procedures exist which are generally very fast and accurate [1, 7, 8, 9]. Approximation functions over selected intervals are successively sampled and then summed. These procedures are best coded in assembler language and are sufficiently detailed to hinder portability. Because of large memory requirements they are restricted to applications where speed is essential.

A strict rejection technique attributed to von Neumann [1, 10] is slow because of poor sampling efficiency of uniform numbers. Bachelor's method utilizes advantages of both composition and rejection techniques by splitting the normal probability density function into two parts which are sampled by rejection [11]. The algorithm compares favorably in speed and memory requirements with other methods. However, calculation of two logarithms and three random numbers is repeated until the rejection process obtains an acceptable number (acceptance rate of 78 percent).

Improvements in the above methods have been made. For instance, rejection sampling has been used to replace sine/cosine calculations in the Box and Muller method [12]. Also, the central limit theorem approach appears to be particularly well suited to highly parallel computers like ILLIAC IV where many random numbers can be simultaneously generated and then summed [13, 19].

8-2 Conditional Bit Generator

We present a new general method of sampling from an arbitrary distribution which is most often faster and more general than the previously discussed methods. We present our method by example. Suppose a normal number is to be computed with $\mu = 0$ and $\sigma = 1$. Further, the range will be restricted to 8σ. Let $SX_1X_2 \cdot X_3X_4 \ldots$ be the binary expansion (S = sign) of the normal number to be generated. The bits of the normal number are generated from high order to low order, for the normal $P(X_1 = 1)$ equals the probability that the entire pseudonormal number lies between $10.000 \ldots _2$ $= 2_{10}$ and $11.111 \ldots _2 = 4_{10}$. Thus $P(X_1 = 1) = 2[\phi(4) - \phi(2)] = 2(0.5000 - 0.4773) = 0.0454$. Care must be taken. If another high-order bit were appended to the left of X_1, then $X_1 = 1$ would also occur when the normal number was between $110.000 \ldots _2 = 6_{10}$ and $111.111 \ldots _2 = 8_{10}$, and so forth; however, for this example, $\phi(8) - \phi(6) \cong 0$. Of course, $P(X_1 = 0) = 1 - P(X_1 = 1) = 0.9546$.

The next step is to examine the conditional probability $P(X_2 = 1 \mid X_1)$. Specifically,

$$P(X_2 = 1 \mid X_1 = 0) = \frac{P(X_1 = 0 \text{ and } X_2 = 1)}{P(X_1 = 0)} \tag{8.2}$$

thus

$$
\begin{aligned}
P(X_2 = 1 \mid X_1 = 0) &= \frac{\phi(01.111 \ldots = 10.000 \ldots) - \phi(01.000 \ldots)}{\phi(01.111 \ldots = 10.000 \ldots) - \phi(00.000 \ldots)} \\[2mm]
&= \frac{2[\phi(2) - \phi(1)]}{2[\phi(2) - \phi(0)]} \\[2mm]
&= \frac{2(0.4773 - 0.3412)}{0.9546} = 0.285
\end{aligned}
\tag{8.3}
$$

Since the 2s cancel from the numerator and denominator of the conditional probability statement, it is only necessary to consider one side of the distribution. In a similar manner,

$$
\begin{aligned}
P(X_2 = 1 \mid X_1 = 1) &= \frac{P(X_1 = 1 \text{ and } X_2 = 1)}{P(X_1 = 1)} \\[2mm]
&= \frac{\phi(11.111 \ldots = 100.000 \ldots) - \phi(11.000 \ldots)}{\phi(11.111 \ldots = 100.000 \ldots) - \phi(10.000 \ldots)} \\[2mm]
&= \frac{\phi(4) - \phi(3)}{\phi(4) - \phi(2)} = 0.0573
\end{aligned}
\tag{8.4}
$$

For selection of the third bit,

$$
\begin{aligned}
P(X_3 = 1 \mid X_1 = 0, X_2 = 0) &= \frac{\phi(00.111 \ldots = 1.0) - \phi(00.100 \ldots)}{\phi(00.111 \ldots = 1.0) - \phi(00.000 \ldots)} \\[2mm]
&= \frac{\phi(1) - \phi(\frac{1}{2})}{\phi(1) - \phi(0)} = 0.439
\end{aligned}
\tag{8.5}
$$

$$
\begin{aligned}
P(X_3 = 1 \mid X_1 = 0, X_2 = 1) & \\[2mm]
&= \frac{\phi(01.111 \ldots = 10.000 \ldots) - \phi(01.100 \ldots)}{\phi(01.111 \ldots = 10.000 \ldots) - \phi(01.000 \ldots)} \\[2mm]
&= \frac{\phi(2) - \phi(1 + \frac{1}{2})}{\phi(2) - \phi(1)} = 0.324
\end{aligned}
\tag{8.6}
$$

$$P(X_3 = 1 \mid X_1 = 1, X_2 = 0)$$

$$= \frac{\phi(10.111 \ldots = 11.000 \ldots) - \phi(10.100 \ldots)}{\phi(10.111 \ldots = 11.000 \ldots) - \phi(10.000 \ldots)}$$

$$= \frac{\phi(3) - \phi(2 + \tfrac{1}{2})}{\phi(3) - \phi(2)} = 0.227 \tag{8.7}$$

and

$$P(X_3 = 1 \mid X_1 = 1, X_2 = 1)$$

$$= \frac{\phi(11.111 \ldots = 100.000 \ldots) - \phi(11.100 \ldots)}{\phi(11.111 \ldots = 100.000 \ldots) - \phi(11.100 \ldots)}$$

$$= \frac{\phi(4) - \phi(3 + \tfrac{1}{2})}{\phi(4) - \phi(3)} = 0.154 \tag{8.8}$$

This process is then continued for the next eight conditional probabilities $P(X_4 = 1 \mid X_1, X_2, X_3)$. Suppose the first j bits of the number are selected and this number is written $\bar{X}_j = X_1 X_2 X_3 X_4 \ldots X_j$. It is inductively apparent from previous computations that the probability of selecting $X_{j+1} = 1$ is

$$P(X_{j+1} = 1 \mid \bar{X}_j) = \frac{\phi(\bar{X}_j + 2^{2-j}) - \phi(\bar{X}_j + 2^{1-j})}{\phi(\bar{X}_j + 2^{2-j}) - \phi(\bar{X}_j)}$$

$$j = 0, 1, 2, \ldots \tag{8.9}$$

and define $P(X_1 = 1 \mid \bar{X}_0) = P(X_1 = 1)$. Substitute $Z_j = \bar{X}_j + 2^{2-j}$ and divide the numerator and denominator by 2^{2-j} and obtain

$$P(X_{j+1} = 1 \mid \bar{X}_j) = 1/2 \frac{[\phi(Z_j) - \phi(Z_j - 2^{1-j})]/2^{1-j}}{[\phi(Z_j) - \phi(Z_j - 2^{2-j})]/2^{2-j}} \tag{8.10}$$

The limit as $j \to \infty$ of Equation (8.10) can be taken (noting the familiar expression for the derivative) as

$$\lim_{j \to \infty} P(X_{j+1} = 1 \mid \bar{X}_j) = \frac{1}{2} \frac{\phi'(Z_\infty)}{\phi'(Z_\infty)} = \frac{1}{2}$$

PAYNE'S THEOREM. If F is a differentiable cumulative probability distribution and a number X is sampled from F, then the probability of occurrence of the jth digit of X is uniformly distributed in the limit as $j \to \infty$.

Proof. The proof is immediate from the above limit consideration.

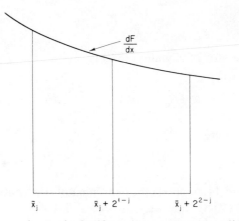

$$\bar{x}_j \qquad \bar{x}_j + 2^{1-j} \qquad \bar{x}_j + 2^{2-j}$$

Figure 8-1. Areas under typical cdf used to compute conditional probability

Graphically, the area under dF/dX between $\bar{X}_j + 2^{1-j}$ and $\bar{X}_j + 2^{2-j}$ is divided by the area under the curve between $\bar{X}_j$ to $\bar{X}_j + 2^{2-j}$. See Figure 8-1. The conditions on F can be reduced to require only continuity of F, and the theorem can be proved with a more complicated proof.

As with most limit theorems, no information is given in the theorem on how large j must be before the distribution of the jth digit is close to uniformity. Fortunately, the number is usually small (particularly for the normal).

For clarity's sake, a normal conditional bit generator will be developed. Extension to any continuous cdf F is immediate. The aim in construction of the generator is to select as few high-order bits as possible with the use of tables and then to attach a random (pseudorandom) number to form the low-order digits. First, however, tables of conditional probabilities must be computed.

8-3 Computation of Conditional Probabilities

Numerical integration of $dF(X)$ must be accomplished to compile tables of $P(X_{j+1} = 1 \mid \bar{X}_j)$ of Equation (8.9). For the normal example, define $R_j = 2^{2-j}$ or recursively,

$$R_j = \frac{R_{j-1}}{2} \qquad R_0 = 4 \tag{8.11}$$

and

$$U_j = \bar{X}_j + R_j \tag{8.12}$$

$$V_j = \bar{X}_j + R_{j+1} \tag{8.13}$$

The numerator of Equation (8.9) is

$$\phi(U_j) - \phi(V_j) = \int_{V_j}^{U_j} d\phi(X) \tag{8.14}$$

Similarly substitute

$$U_j' = U_j, \qquad V_j' = \bar{X}_j \tag{8.15}$$

into Equation (8.14) to obtain the denominator. Integration over the largest interval occurs first, and as j increases, integration is computed over smaller intervals, thus maintaining accuracy.

The next set of probabilities, $P(X_{j+2} = 1 \mid \bar{X}_{j+1})$, can be computed by replacement:

$$R_{j+1} \leftarrow \frac{R_j}{2} \tag{8.16}$$

$$U_{j+1} \leftarrow \bar{X}_{j+1} + R_{j+1} \tag{8.17}$$

and

$$V_{j+1} \leftarrow \bar{X}_{j+1} + R_{j+2}. \tag{8.18}$$

All possible permutations of the bits of $\bar{X}_j$ must be enumerated in order while maintaining the correct binary point. Let there be k bits to the left of the binary point. All possible 2^j combinations of 1s and 0s of $\bar{X}_j$ can be computed by evaluation of

$$\frac{i}{2^{j-k}} \qquad i = 0, 1, \ldots, 2^{j-1} \tag{8.19}$$

in floating point. The table of $P(X_{j+1} = 1 \mid \bar{X}_j), j = 0 \ldots, L - 1$, consists of $2^{L+1} - 1$ conditional probabilities. A FORTRAN subprogram which generates this table is presented (see Figure 8-2), and it calls the numerical integration routine ROMINT. The quadrature scheme is algorithm 351, an improved Romberg technique [15].

The variable names in this subprogram closely resemble the notation used in the text and in algorithm 351. Complete tables of $P(X_{j+1} = 1 \mid \bar{X}_j)$ are given in Figure 8-3 up to $j = 4$, and selected values are given for $j = 5, 6, 7$, and 8.

The table was also computed using the IBM FORTRAN-supplied error

```
1          SUBROUTINE PTABLE(L,K,TABLE,BIGERR,ERR,EPS,N,MAXE)
2  C ROMBERG QUADRATURE, COMM. ACM, 12, 6, (JUNE 1969), 324.
3  C 6=TOTAL NUMBER OF BITS SAMPLED USING TABLE.
4  C K=NUMBER OF BITS TO THE LEFT OF BINARY POINT.
5  C BIGERR=-1.=ERROR IN ROMINT.
6  C ERR,EPS,N,MAXE=ROMINT PARAMETERS.
7  C R=2**(K-J) IN ITERATIVE FORM.
8  C J=BIT COUNTER, COUNTS TO L.
9  C J2=2**J IN ITERATIVE FORM.
10         DOUBLE PRECISION TOP,BOT,ERR,EPS,UJ,VJ,XBAR,P,R
11         DIMENSION TABLE(1)
12         BIGERR=0
13         R=2**(K+1)
14         J2=1
15         DO 1 JJ=1,L
16         J=JJ-1
17         R=R/2
18 C II=PERMUTATION COUNTER.
19         DO 2 II=1,J2
20 C OBTAIN BIT COMBINATION IN INTEGER FORM.
21         I=II-1
22 C NORMALIZE BIT COMBINATION TO GET XBAR(J).
23         XBAR=I*R
24 C CALCULATE UPPER AND LOWER LIMITS.
25         UJ=XBAR+R
26         VJ=XBAR+R/2
27 C INTEGRATE NUMERATOR, THEN DENOMINATOR.
28         MAXE1=MAXE
29 C ROMINT IS ALGORITHM 351, FAIRWEATHER, G. MODIFIED.
30         CALL ROMINT(TOP,ERR,EPS,VJ,UJ,N,MAXE1)
31         IF(MAXE1.EQ.0.OR.TOP.LT.0) GO TO 3
32         MAXE2=MAXE
33         CALL ROMINT(BOT,ERR,EPS,XBAR,UJ,N,MAXE2)
34         IF(MAXE2.EQ.0.OR.BOT.LT.0) GO TO 3
35 C CALCULATE TABLE ENTRY
36         TABLE(I+J2)=TOP/BOT
37         GO TO 2
38 3       TABLE(I+J2)=-1.
39         BIGERR=-1.
40 2       CONTINUE
41         J2=2*J2
42 1       CONTINUE
43         RETURN
44         END
```

Figure 8-2. FORTRAN subprogram PTABLE to computer conditional probabilities

function subprogram, DERF. Double-precision results agreed to 8 decimal digits, and total computation time on an IBM 360/67 for both methods was less than 10 seconds. From the table, the distribution of the ninth bit of the normal number is quite close to 1/2, but even for 6 bits, the maximum deviation from 1/2 is not great.

If three integer bits are computed (pseudonormal number between $\pm 8\sigma$, the probability that the high-order bit equals 1 was 0.00006334. The table presented here (2 integer bits) differed at most by 0.0003 from corresponding entries in the 3-integer-bit table. Although 3 or more integer bits can be sampled, this would be done only for specialized applications.

XBAR(0)	P(X(1)=1|XBAR(0))
0.0	0.04543980

XBAR(1)	P(X(2)=1|XBAR(1))
0.0	0.28476858
2.00000	0.05797400

XBAR(2)	P(X(3)=1|XBAR(2))
0.0	0.43907970
1.00000	0.32420421
2.00000	0.22708505
3.00000	0.15228540

XBAR(3)	P(X(4)=1|XBAR(3))
0.0	0.48445958
0.50000	0.45350313
1.00000	0.42290229
1.50000	0.39288020
2.00000	0.36364293
2.50000	0.33537388
3.00000	0.30823034
3.50000	0.28234071

XBAR(4)	P(X(5)=1|XBAR(4))
0.0	0.49609876
0.25000	0.48829854
0.50000	0.48050398
0.75000	0.47271895
1.00000	0.46494716
1.25000	0.45719236
1.50000	0.44945818
1.75000	0.44174832
2.00000	0.43406636
2.25000	0.42641592
2.50000	0.41880035
2.75000	0.41122317
3.00000	0.40368766
3.25000	0.39619714
3.50000	0.38875473
3.75000	0.38136363

XBAR(5)	P(X(6)=1|XBAR(5))
0.0	0.49902374
0.12500	0.49707127
0.25000	0.49511892
0.37500	0.49316669
0.50000	0.49121469
0.62500	0.48926294
0.75000	0.48731154
0.87500	0.48536050
1.00000	0.48340988
1.12500	0.48145986
1.25000	0.47951031
.	.
.	.
.	.

3.50000	0.44458240
3.62500	0.44265473
3.75000	0.44072872
3.87500	0.43880457

XBAR(6)	P(X(7)=1|XBAR(6))
0.0	0.49975586
0.06250	0.49926764
0.12500	0.49877936
0.18750	0.49829113
0.25000	0.49780291
0.31250	0.49731469
0.37500	0.49682647
.	.
.	.
.	.
3.75000	0.47049564
3.81250	0.47000915
3.87500	0.46952266
3.93750	0.46903628

XBAR(7)	P(X(8)=1|XBAR(7))
0.0	0.49993896
0.03125	0.49981689
0.06250	0.49969482
0.09375	0.49957275
0.12500	0.49945068
.	.
.	.
.	.
3.87500	0.48480719
3.90625	0.48468524
3.93750	0.48456329
3.96875	0.48444134

XBAR(8)	P(X(9)=1|XBAR(8))
0.0	0.49998474
0.01563	0.49995422
0.03125	0.49992371
0.04688	0.49989319
0.06250	0.49986267
0.07813	0.49983215
0.09375	0.49980164
0.10938	0.49977112
0.12500	0.49974060
.	.
.	.
.	.
3.92188	0.49232543
3.93750	0.49229491
3.95313	0.49226445
3.96875	0.49223393
3.98438	0.49220341

Figure 8-3. Table values for $K = 2$ and $L = 9$ (Normal generator)

8-4 Normal Number Generator

A random number, U, and a comparison are used to sample each bit. If

$$U \le P(X_{j+1} = 1 \mid \overline{X}_j) \qquad (8.20)$$

is true, a 1 bit is generated for X_{j+1} and a 0 bit if the inequality is false. The generated bits are used to form table index m. The mapping function to retrieve $P(X_{j+1} = 1 \mid \overline{X}_j)$ can be written in terms of $\overline{X}_j$ as

$$m = \overline{X}_j + 2^{j-1} \qquad j = 1, \ldots, L \qquad (8.21)$$

that is, m is the address of $P(X_{j+1} = 1 \mid \overline{X}_j)$ in the table. When as many bits as necessary have been sampled using the conditional probability table, a uniform number W is appended to the low-order positions to form a full-length floating-point number C. This results when

$$C = \frac{\overline{X}_L + W}{2^{L-k}} \qquad (8.22)$$

is evaluated in floating point (L bits generated from the table; k integer bits). With use of a quality pseudorandom number generator [16, 17] called RAND a generator used to produce pseudonormal numbers can be easily coded in FORTRAN. A possible coding is shown in Figure 8-4.

High-quality pseudonormal numbers could be obtained by sampling 6 bits using the conditional probability table (63 entries) and then appending a pseudorandom number to form the low-order bits.

8-5 Discussion

Conditional bit sampling must be considered one of the major methods of sampling from continuous distributions. The advantages are as follows.

(1) Most arbitrary continuous functions can be sampled.

(2) Arbitrary quality of the sampled numbers is determined by the number of bits computed from a comparison of a uniform number and the entries in a conditional probability table. The less quality desired, the smaller the table of conditional probabilities, and the faster the generator can be executed.

(3) The algorithm can be coded in portable compiler languages. If extreme speed is desired, we recommend that the table of conditional probabilities be computed by a compiler program. These table constants are then passed as assembler constants along with a macrocall to the assembler [18]. A conditional sampling macro would be expanded by the assembler into a highly efficient sampling routine with arbitrary accuracy. Examples of such conditional assembler macro techniques can be found in

```
 1      FUNCTION CONBIT(L,K,TABLE)
 2      DIMENSION TABLE(1)
 3      EQUIVALENCE (IW,W)
 4    C FCT CONBIT RETURNS A NORMAL DEVIATE BY BIT SAMPLING
 5    C USING TABLE(M) OF CONDITIONAL PROBABILITIES, (SEE M BELOW)
 6    C L=TOTAL NUMBER OF BITS SAMPLED USING TABLE.
 7    C K=NUMBER OF BITS TO THE LEFT OF BINARY POINT.
 8    C RAND=REAL PSEUDORANDOM NUMBER ON (0,1).
 9    C J=BIT COUNTER.
10    C I=INTEGER FORM OF CONBIT.
11    C M=MAPPING OF TABLE ADDRESS FOR P(X(J+1)=1|XBAR(J)).
12    C W=IW=UNIFORM PSEUDORANDOM NUMBER APPENDED TO TABLE
13    C GENERATED BITS. LOW ORDER BIT OF W USED FOR SIGN.
14      I=0
15      DO 10 J=1,L
16      M=I+2**(J-1)
17      I=2*I
18      IF(RAND(NO).LE.TABLE(M)) I=I+1
19    C TEST FOR 1 OR 0 IN JTH BIT POSITION COMPLETED.
20 10   CONTINUE
21      W=RAND(NO)
22      SIGN=2*(IW-2*(IW/2))-1.0
23      CONBIT=SIGN*(I+W)/(2.0**(L-K))
24      RETURN
25      END
```

Figure 8-4. FORTRAN subroutine for conditional bit sampling

[19] (see macro-generated subroutine QKRSRT, p. 314). It is possible to partition a pseudorandom number so that a single pseudorandom number can be used for several table comparisons [20].

(4) Conditional bit sampling can be adapted to sample discrete distributions. Since only a finite number of bits are generated in these cases, the algorithm must be modified slightly. When the number of finite outcomes is great, the method can prove practical.

8-6 Sampling with CBS Generator

Binomial and Hypergeometric

Any summable frequency function whose random variables take on a finite number of values can be sampled with ICONBT. Scaling for fractional-valued variables must be done on integer output from ICONBT (the integer equivalent of CONBIT given in Section 8-4).

Logical operation techniques [21] for generating pseudobinomial numbers and Davis's method [22] become lengthy when n is large or when p is small. Binary-directed table search becomes impractical when n is very large. Conditional bit sampling is relatively unaffected by either large n or small p: arbitrary accuracy (number of high-order bits) is determined by the number of bits sampled from table comparisons. Accuracy can be sacrificed when np is large (on low-order bits) by appending a uniform pseudorandom number [$P(X_{j+1} = 1 \mid \bar{X}_j) \equiv 1/2$ for large j for most $\bar{X}_j$] to the right of the table-generated high-order bits. Sampling speed is thus increased.

The author knows of no method, save binary-directed table search, for sampling from the hypergeometric distribution.

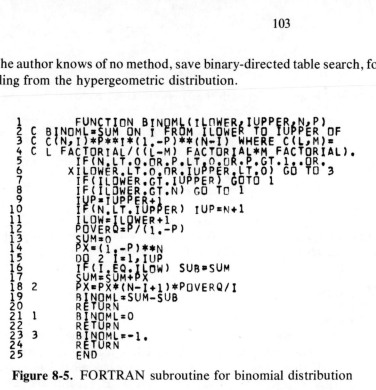

```
 1         FUNCTION BINOML(ILOWER,IUPPER,N,P)
 2 C BINOML=SUM ON I FROM ILOWER TO IUPPER OF
 3 C C(N,I)*P**I*(1.-P)**(N-I) WHERE C(L,M)=
 4 C L FACTORIAL/((L-M) FACTORIAL*M FACTORIAL).
 5         IF(N.LT.0.OR.P.LT.0.OR.P.GT.1..OR.
 6       XILOWER.LT.0.OR.IUPPER.LT.0) GO TO 3
 7         IF(ILOWER.GT.IUPPER) GOTO 1
 8         IF(ILOWER.GT.N) GO TO 1
 9         IUP=IUPPER+1
10         IF(N.LT.IUPPER) IUP=N+1
11         ILOW=ILOWER+1
12         POVERQ=P/(1.-P)
13         SUM=0
14         PX=(1.-P)**N
15         DO 2 I=1,IUP
16         IF(I.EQ.ILOW) SUB=SUM
17         SUM=SUM+PX
18 2       PX=PX*(N-I+1)*POVERQ/I
19         BINOML=SUM-SUB
20         RETURN
21 1       BINOML=0
22         RETURN
23 3       BINOML=-1.
24         RETURN
25         END
```

Figure 8-5. FORTRAN subroutine for binomial distribution

```
 1         FUNCTION HYPERG(ILOWER,IUPPER,N,J,K)
 2 C HYPERG=SUM ON I FROM ILOWER TO IUPPER OF
 3 C C(K,I)*C(N-J,K-I)/C(N,K) WHERE C(L,M)=
 4 C L FACTORIAL/((L-M) FACTORIAL*M FACTORIAL).
 5         IF(ILOWER.LT.0.OR.IUPPER.LT.0.OR.N.LT.0
 6       X.OR.J.LT.0.OR.K.LT.0.OR.J.GT.N.OR.K.GT.N)
 7       XGO TO 5
 8         LEAST=J
 9         IF(J.GT.K) LEAST=K
10         IF(ILOWER.GT.IUPPER.OR.ILOWER.GT.LEAST)
11       XGO TO 3
12         IUP=IUPPER+1
13         ILOW=ILOWER+1
14         IF(IUPPER.GT.LEAST) IUP=LEAST+1
15         PX=1.
16         DO 1 I=1,J
17 1       PX=PX*(N-K-I+1)/(N-I+1)
18         SUM=0
19         SUB=0
20         IBEGIN=0
21 4       IBEGIN=IBEGIN+1
22         IF(N-J-K+IBEGIN.LE.0) GO TO 4
23         IF(IBEGIN.GT.IUP) GO TO 3
24         DO 2 I=IBEGIN,IUP
25         IF(I.EQ.ILOW) SUB=SUM
26         SUM=SUM+PX
27 2       PX=PX*(J-I+1)*(K-I+1)/(I*(N-J-K+I))
28         HYPERG=SUM-SUB
29         RETURN
30 3       HYPERG=0
31         RETURN
32 5       HYPERG=-1.
33         RETURN
34         END
```

Figure 8-6. FORTRAN subroutine for hypergeometric distribution

Poisson and Negative Binomial

Numbers sampled from both the Poisson, with frequency function

$$f_1(i) = \frac{e^{-\lambda}\lambda^i}{i} \qquad \lambda > 0, \, i = 0, 1, 2, \ldots \qquad (8.23)$$

and the negative binomial, with frequency function

$$f_2(i) = \binom{k + i - 1}{i} p^k (1 - p)^i \qquad (8.24)$$

where $k > 0$, $0 \le p \le 1$, and $i = 0, 1, \ldots$, take on the denumerable number of integer values $0, 1, 2, \ldots$. For both distributions $(m = 1, 2)$

$$\sum_{i=0}^{Y'_N} f_m(i) \cong 1 \qquad (8.25)$$

for large integer Y'_N.

Values of Y'_N depend upon λ for Poisson binomial and upon k and p for the negative binomial: practical values of Y'_N are usually small, less than 40 for $\lambda \le 20$ (Poisson) and $k \le 45$, $p \le \frac{1}{2}$ (negative binomial). Let $X_1 X_2 X_3 \ldots X_N$ be the N-bit binary expansion of integer $X'_N \le Y'_N$. Bits are generated from high order to low order in subprogram ICONBT with arguments L equal to the number of bits to the left of the binary point and single-dimensioned array TABLE, a table of one probability and $2^L - 2$ conditional probabilities, $P(X_{i+1} = 1 \mid \bar{X}_i)$, $\bar{X}_i = X_1 X_2 \ldots X_i$, $i = 1, 2, \ldots, N - 1$, that is, $(\bar{X}_i)_{10} = X_1 2^{N-1} + X_2 2^{N-2} + \ldots + X_i 2^{N-i}$. Fixed point function subprogram ICONBT returns L = bit pseudo-Poisson numbers. ICONBT calls pseudorandom number generator, RAND, which returns a floating-point number on $(0, 1)$. Either a maximum-period primitive root Lehmer (see Part 1) or a shift register sequence, pseudorandom generator, is recommended. The array TABLE, passed in the argument list of ICONBT, is a table beginning with a probability followed by conditional probabilities which must be stored in the order $P(X_1 = 1)$, $P(X_2 = 1 \mid X_1 = 0)$, $P(X_2 = 1 \mid X_1 = 1)$, $P(X_3 = 1 \mid X_1 = 0, X_2 = 0)$, $P(X_3 = 1 \mid X_1 = 0, X_2 = 1)$, $P(X_3 = 1 \mid X_1 = 1, X_2 = 0)$, $\ldots$, $P(X_N = 1 \mid X_1 = 1, X_2 = 1, \ldots, X_{N-1} = 1)$.

Since the negative binomial has different parameters than the Poisson, the first line in subroutine IPTABL must be altered to SUBROUTINE IPTABL(L, TABLE, ERR, K, P), and lines 22 and 23 must be changed to TOP=GINEG(ILOW, IHI, K, P) and BOT=BINEG(ILOW, IHI, KM, P), respectively, to generate the table of conditional probabilities (and one probability necessary for conditional bit sampling the negative binomial) (see PTABLE in Section 8-2).

For the Poisson, λ pseudorandom numbers need to be generated, on the average, using the traditional multiplication technique [22]. As λ (about $\lambda = 5$) increases, conditional bit sampling surpasses this traditional technique with respect to speed.

Algorithm 342 [23] obtains pseudo-Poisson-distributed numbers from a binary-directed table search on the ordinate of the cumulative Poisson distribution. Although algorithm 342 is considerably longer than ICONBT, it generates pseudo-Poisson variates with greater speed for small (less than 20) λ. This results from the fact that one pseudorandom number is generated for each bit generated from table comparisons in this algorithm, while only one pseudorandom number is needed along with a table search in ACM's algorithm 342. If λ is large, then this generator may have the advantage that only a specified number of high-order bits need be generated for rough approximation to pseudo-Poisson numbers.

No known technique, other than table search, for sampling from the negative binomial is known to the author.

Extensions to sampling from any distributions taking on a denumerable number of values is immediate. If fractional values are assumed by the random variable, scaling must be implemented outside ICONBT since only integer values are returned.

Gamma and Beta

Numbers sampled from both the gamma distribution, with density

$$f_5(x) = \frac{x^{p-1}e^{-x/b}}{\Gamma(p)b^p} \tag{8.26}$$

where $p \geq 1$, $b > 0$, and $0 \leq x \leq \infty$ [for $p = 1$, define $f_5(x) = b^{-1}e^{-x/b}$], and the beta distribution, with density

$$f_6(x) = \frac{\Gamma(a + b)x^{a-1}(1 - x)^{b-1}}{\Gamma(a)(b)} \tag{8.27}$$

where $a > 1$, $b > 1$, and $0 \leq x \leq 1$, take on positive real values.

Subroutine PTABLE presented in Section 8-2 can be used to generate the single probability and remaining conditional probabilities necessary for conditional bit sampling from the gamma and beta distributions. Line 1 in subroutine PTABLE must be changed to SUBROUTINE PTABLE(L, K, TABLE, BIGERR, ERR, EPS, N, MAXE, P, B) or SUBROUTINE PTABLE(L, K, TABLE BIGERR, ERR, EPS, N, MAXE, A, B). Line 30 of subroutine PTABLE must be changed to CALL ROMINT(TOP, ERR,

```
 1          FUNCTION ICONBT(L,TABLE)
 2 C L=TOTAL NUMBER OF BITS SAMPLED USING TABLE.
 3 C RAND=REAL PSEUDORANDOM NUMBER ON (0,1).
 4 C J=BIT COUNTER.
 5 C I=INTEGER FORM OF CONBIT.
 6          DIMENSION TABLE(1)
 7          K=1
 8          I=0
 9          DO 1 J=1,L
10          M=I+K
11          I=2*I
12 C TEST FOR 1 OR 0 IN J-TH BIT POSITION.
13          IF(RAND(NO).LE.TABLE(M)) I=I+1
14          K=2*K
15 1        CONTINUE
16          ICONBT=I
17          RETURN
18          END
```

Figure 8-7. FORTRAN subroutine for sampling a discrete pdf

```
 1          FUNCTION BINEG(ILOWER,IUPPER,K,P)
 2 C BINEG=SUM ON I FROM ILOWER TO IUPPER OF
 3 C C(K+I-1,I)*P**K*(1.-P)**I WHERE C(L,M)=
 4 C L FACTORIAL/((L-M) FACTORIAL*M FACTORIAL).
 5          IF(ILOWER.LT.0.OR.IUPPER.LT.0.OR.K.LT.1
 6      X.OR.P.LT.0.OR.P.GT.1.) GO TO 3
 7          IF(ILOWER.GT.IUPPER) GO TO 1
 8          ILOW=ILOWER+1
 9          IUP=IUPPER+1
10          Q=1.-P
11          SUM=0
12          PX=P**K
13          DO 2 I=1,IUP
14          IF(I.EQ.ILOW) SUB=SUM
15          SUM=SUM+PX
16 2        PX=PX*(K+I-1)*Q/I
17          BINEG=SUM-SUB
18          RETURN
19 1        BINEG=0
20          RETURN
21 3        BINEG=-1.
22          RETURN
23          END
```

Figure 8-8. FORTRAN subroutine for negative binomial distribution

EPS, VJ, UJ, N, MAXEL, P, B) or CALL ROMINT(TOP, ERR, EPS, VJ, UJ, N, MAXEL, A, B), and line 33 must be changed to CALL ROMINT(BOT, ERR, EPS, XBAR, UJ, N, MAXE 2, A, B), respectively, for the gamma or beta distributions.

Algorithm 351, subroutine ROMINT, was used to perform integrations on both the gamma and the beta distributions. Unfortunately, no easy way exists in FORTRAN to pass to subroutine ROMINT the name and parameters of the function to be integrated. Function F in ROMINT (algorithm 351) must be replaced (lines 6, 9, and 41) by the appropriate GAMMAD(X, P, B1) or BETA(X, A1, B1) function subprograms which define the gamma and beta integrands for ROMINT. If ROMINT fails (MAXE = 0) to achieve desired accuracy (EPS), then the appropriate value

```
 1          FUNCTION BETA(X,A,B)
 2          DOUBLE PRECISION BETA,X,A,B
 3          IF(X.LT.0D0.OR.X.GT.1.D0.OR .A.LE.1.D0
 4         X.OR.B.LE.1.D0) GO TO 1
 5          BETA=X**(A-1)*(1.D0-X)**(B-1)*DGAMMA(A+B)
 6         X/(DGAMMA(A)*DGAMMA(B))
 7          RETURN
 8  1       BETA=-1.
 9          RETURN
10          END
```

Figure 8-9. FORTRAN subroutine for beta distribution

```
 1          FUNCTION GAMMAD(X,P,B)
 2          DOUBLE PRECISION GAMMAD,X,P,B
 3          IF(X.LT.0D0.OR.P.LT.1.D0.OR.B.LT.0D0) GO TO 1
 4          IF(X.EQ.0D0.AND.P.EQ.1.D0) GO TO 2
 5          GAMMAD=X**(P-1.D0)*DEXP(-X/B)/(B**P*DGAMMA(P))
 6          RETURN
 7  2       GAMMAD=1.D0/B
 8          RETURN
 9  1       GAMMAD=-1.
10          RETURN
11          END
```

Figure 8-10. FORTRAN subroutine for gamma distribution

```
 1          FUNCTION PCONBT(L,K,TABLE)
 2  C FCT PCONBT RETURNS A PSEUDORANDOM VARIATE BY BIT SAMPLING
 3  C USING TABLE(M) OF CONDITIONAL PROBABILITIES
 4  C L=TOTAL NUMBER OF BITS SAMPLED USING TABLE.
 5  C K=NUMBER OF BITS TO THE LEFT OF BINARY POINT.
 6  C RAND=REAL PSEUDORANDOM NUMBER ON (0,1)
 7  C J=BIT COUNTER, COUNTS TO L.
 8  C I=INTEGER FORM OF PCONBT.
 9  C M=MAPPING OF TABLE ADDRESS FOR P(X(J+1)=1|XBAR(J)).
10  C J2=2**(J-1) IN ITERATIVE FORM.
11          DIMENSION TABLE(1)
12          I=0
13          J2=1
14          DO 1 J=1,L
15          M=I+J2
16          I=2*I
17  C TEST FOR 1 OR 0 IN J-TH BIT POSITION.
18          IF(RAND(N0).LE.TABLE(M)) I=I+1
19          J2=2*J2
20  1       CONTINUE
21          PCONBT=(I+RAND(N0))/(2.0**(L-K))
22          RETURN
23          END
```

Figure 8-11. FORTRAN subroutine for continuous distribution sampling

in TABLE is set to -1, and -1 is returned at BIGERR. Other quadrature schemes should be tried for the troublesome integrand.

Function subprogram PCONBT samples from either the gamma or the beta distribution depending upon the table of conditional probabilities used for table comparisons.

If variates take on only positive real values, PCONBT can be used to sample, with arbitrary accuracy of high- or low-order digits, any continuous probability distribution function (cdf) whose variates take on nonnegative real values.

If the distribution to be sampled is symmetrically distributed about zero, CONBIT of Section 8-1 is an appropriate sampling routine. For asymmetric distributed random variables about zero, the sign may be treated as another bit, but subroutine PTABLE must be modified to construct the conditional probability table for both positive and negative portions of the distribution.

All the generators in this section illustrate by example[a] conditional bit sampling for a wide class of probability distributions. Although sampling methods specific to some distributions prove faster and more efficient than conditional bit sampling, no other general sampling method can simultaneously provide arbitrary accuracy, portability, moderate memory requirements, and, often, speed.

References

[1] Muller, N.E. A comparison of methods for generating normal deviates on digital computers. *J. ACM*, 6: 376-383 (July 1959).

[2] Box, G. and N. Muller. A note on the generation of normal deviates. *Ann. Math. Statis.*: 28 (1958), 610.

[3] Kronmal, R. Evaluation of a pseudo-random normal number generator. *J. ACM*, 11(3): 357-363 (July 1964).

[4] Muller, M.E. Generation of normal deviates. *Tech. Rept. No. 13, Statis. Tech. Res. Group*, Dept. of Math., Princeton University, 1957.

[5] Hastings, C. "Approximations for digital computations." Princeton, N.J.: Princeton University Press, 1955, p. 192.

[6] Teichroew, D. Distribution sampling with high speed computers. Ph.D. Thesis, North Carolina State College, 1953.

[7] Marsaglia, G., and T.A. Bray. A convenient method for generating normal variables. *Comm. ACM*, 6(3): 260-264 (July 1964).

[8] Marsaglia, G., MacLaren, M.P., and Bray, T.A. A fast procedure for generating normal random variables. *Comm. ACM*, 7(1): 4-10 (Jan. 1964).

[9] Naylor, T.H., Balintfy, J.L., Burdick, D.S., and Chu, K. "Computer Simulation Techniques." New York: Wiley, 1966.

[10] von Neumann, J. Various techniques used in connection with random digits. *Natl. Bur. Std. Appl. Math.*, 36, Sept. 12, 1959.

[11] Tocher, K.D. "The Art of Simulation." London: The English University Press, 1963, p. 27.

---

[a]The author is indebted to Prof. W. H. Payne of Washington State University for these programs and algorithms.

[12] Bell, J.R. Algorithm 334—Normal random deviates. *Comm. ACM*, 11(7): 498 (July 1968).

[13] Kuck, D.J. ILLIAC IV software and applications programming. *IEEE Trans. on Computers*, C17(8): 758-770 (Aug. 1968).

[14] Winje, G.L. Random nunber generators for ILLIAC IV. Dept. of Computer Science, University of Illinois at Urbana. June 1969.

[15] Fairweather, G. Algorithm 351—Modified Romberg quadrature. *Comm. ACM*, 12(6): 324-325 (June 1969).

[16] Payne, W., Rabung, J.R., and Bogyo, T. Coding the Lehmer pseudo-random number generator. *Comm. ACM*, 12(2): 85-86 (Feb. 1969).

[17] Whittlesey, J.R.B. A comparison of correlation behavior of random number generators for the IBM 360. *Comm. ACM*, 11(9): 641-655 (Sept. 1968).

[18] Lewis, T.G. Microprogramming probability distribution sampling. *Proc. 25th Nat. ACM Conf.*, August 1972.

[19] Payne, W.H. "Machine, Assembly, and Systems Programming for the IBM 360." New York: Harper & Row, 1969.

[20] Donnelly, T. Some techniques for using pseudorandom numbers in computer simulation. *Comm. ACM*, 12(7): 392-394 (July 1969).

[21] Donnelly, T. Some techniques for using pseudorandom numbers in computer simulation. *Comm. ACM*, 12(7): 392-394 (July 1969).

[22] Tocher, K.D. "The Art of Simulation." London: The English University Press, 1963, p. 40.

[23] Snow, R.H. Algorithm 342: Generator of random numbers satisfying the Poisson distribution. *Comm. ACM*, 11(12): 819-820 (Dec. 1968).

Appendixes for Generators

Comparison of the GFSR algorithm: $X^{98} + X^{27} + 1$, using delayed columns equal 9800. The algorithm results for (*a*) 15-bit Interdata 4 (simulated), (*b*) 31-bit IBM 360, (*c*) 35-bit SRU 1108.

0.36964	0.3696329	0.36963297
0.40632	0.4063136	0.40631372
0.42878	0.4287784	0.42877845
0.47410	0.4741138	0.47411389
0.95318	0.9531578	0.95315778
0.77388	0.7738667	0.77386667
0.10315	0.1031532	0.10315326
0.55989	0.5598878	0.55988777
(*a*)	(*b*)	(*c*)

Appendix B:
Program for HD1
(IBM 360/65)

```
RAND        CSECT
*
*           LEHMER CONGRUENTIAL MIXED PSEUDORANDOM NUMBER GENERATOR
*           CONTRIBUTED BY L. RICHARD TURNER, NASA LEWIS RES. CTR.,
*           CLEVELAND, OHIO.
*
            USING RAND,15
            L     1,OLD
            M     0,MULT
            BCTR  1,0
            ST    1,OLD
            SUR   0,0
            AD    0,FLOT
            BR    14
            DS    0D
FLOT        DC    X'46000000'
OLD         DC    F'1'
MULT        DC    X'A21FA361'
            END
```

Courtesy of Richard Turner, Lewis Research Center (NASA), Cleveland, Ohio.

Appendix C:
Program for HD2
(IBM 360/65)

```
RAND      CSECT
*
*         LEHMER CONGRUENTIAL MULTIPLICATIVE PSEUDORANDOM NUMBER GEN
*         CONTRIBUTED BY W. H. PAYNE, WASHINGTON STATE U., PULLMAN, WA.
*
          PRINT DOUBLE
          L       1,0(,1)         GET ADD OF NUM IN REG 1
          ST      1,24(,13)       STORE ADD OF NUM IN SAVE AREA
          USING RAND,15
          SPM     0
          L       1,=F'630360016'   K=14**29=630360016 MOD(2**31-1)
          M       0,XD              K*X(N)
          SLDA    0,1
          SRL     1,1
          AR      0,1             X'(N+1)=K*X(N)+R  (LOW ORDER 31 BITS)
          BC      1,OVF
CONT      ST      0,XD
          L       1,24(,13)       GET ADD OF NUM FROM SAVE AREA
          ST      0,0(,1)         X'(N+1) IN NUM
          SR      1,1
          SRDA    0,7
          A       0,=X'40000000'
          STM     0,1,TEMP
          SDR     0,0
          AD      0,TEMP
          BR      14
OVF       SLL     0,1
          SRL     0,1
          A       0,=F'1'
          B       CONT
XD        DC      F'5242287'
TEMP      DS      D
          END
```

Courtesy of Prof. W. H. Payne of Washington State University.

Appendix D:
Program for 16 HD2
(IBM 360/65)

```
STMT    SOURCE STATEMENT

    1 RAND     CSECT
    2          SAVE   (14,12)
    3+         CS     0H
    4+         STM    14,12,12(13) SAVE REGISTERS
    5          BALR   15,0
    6          LSING  *,15
    7          L      3,=A(PR)              GET ADDR OF PRIMITIV
    8          L      4,=A(XO)              GET ADDR OF PREVIOUS
    9          L      1,0(,1)       GET ADDR. OF XO LIST.
   10          ST     1,24(,13)     SAVE IN FORTRAN AREA.
   11          L      2,KOUNT       GET GENERATOR COUNT.
   12          S      2,=F'4'       CECREMENT.
   13          ST     2,KOUNT       SAVE GEN. COUNT.
   14          LTR    2,2           TEST IF LIST OF PRIMITIVE
   15          BM     RESET         SKIP TO RESET GEN. COUNT.
   16 GO       L      1,0(3,2)              GET PRIMITIVE ROOT.
   17          M      0,0(4,2)              K*X(N).
   18          SLDA   0,1           ISOLATE HIGH-ORDER 31 BITS
   19          SRL    1,1           ISOLATE LOW-ORDER 31 BITS.
   20          AR     0,1           X(N+1)=K*X(N)+R.
   21          BC     OVF           X(N+1) OVERFLOW.
   22 CONT     ST     0,0(4,2)              X(N+1) REPLACES X(N)
   23          L      1,24(,13)     GET ADDR. OF XO LIST.
   24          ST     0,0(2,1)      X(N+1) IN XO.
   25          ST     0,0(2,4)      STORE IN XO.
   26          SR     1,1           GET LOW-ORDER DOUBLEWORD.
   27          SRDA   0,7           MAKE ROOM FOR MASK.
   28          A      0,=X'40000000' MASK.
   29          STM    0,1,TEMP      SAVE RESULT FOR LATER ADD.
   30          SDR    0,0           CLEAR LOW-ORDER REG..
   31          AC     0,TEMP        PLACE NORMALIZED RESULT IN
   32          B      DONE
   33 OVF      SLL    0,1           SIGN +.
   34          SRL    0,1           SIGN+.
   35          A      0,=F'1'       ADD 1 TO GET X(N+1)+1.
   36          B      CONT
   37 RESET    LA     2,60(,0)
   38          ST     2,KOUNT       RESET GEN. COUNT.
   39          B      GO
   40 DONE     RETURN               (14,12)
   41+DONE     CS     0H
   42+         LM     14,12,12(13) RESTORE THE REGISTERS
   43+         BR     14 RETURN
   44 TEMP     CS     D                     TEMP LOCATION.
   45 KOUNT    CC     F'64'         GENERATOR NUMBER.
   46 XO       CC     F'524287'
   47          CC     F'524287'
   48          CC     F'524287'
   49          CC     F'524287'
   50          CC     F'524287'
   51          CC     F'524287'
   52          CC     F'524287'
   53          CC     F'524287'
   54          CC     F'524287'
   55          CC     F'524287'
```

119

```
56          DC      F'524287'
57          DC      F'524287'
58          DC      F'524287'
59          DC      F'524287'
60          DC      F'524287'
61          DC      F'524287'
62  PR      DC      F'16807'        7**5
63          DC      F'630360016'    14**29
64          DC      F'28629151'     31**5
65          DC      F'184528125'    45**5
66          DC      F'1912518406'   11**13
67          DC      F'1644645313'   39**13
68          DC      F'794991314'    53**13
69          DC      F'2125924067'   88**13
70          DC      F'988683283'    95**13
71          DC      F'727452832'    14**17
72          DC      F'573186566'    22**17
73          DC      F'1141672104'   44**17
74          DC      F'1303680654'   45**17
75          DC      F'376740290'    51**17
76          DC      F'1848369271'   53**17
77          DC      F'1963413634'   62**17
78          LTORG
79                  =A(PR)
80                  =A(XC)
81                  =F'4'
82                  =X'40000000'
83                  =F'1'
84          END
```

Appendix E:
Program for GFSR

```
RAND        CSECT
            PRINT DOUBLE
*
*           GENERALIZED FEEDBACK SHIFT REGISTER PSEUDORANDOM NUM. GENERATOR
*           USING THE PRIMITIVE POLYNOMIAL X**98 + X**27 + 1
*           GENERATES A NORMALIZED RANDOM NUM. IN THE INTERVAL (0,1)
*           REFERENCES :        LEWIS,T.G. AND PAYNE,W.H., GENERALIZED
*                   FEEDBACK SHIFT REGISTER PSEUDORANDOM NUMBER ALGORITHM,
*                   JOURNAL OF THE ACM,VOL. 20, NO. 3, JULY 1973, PP 456-468.
*           PROGRAMMER:  WARREN V. CAMP, U.S.L.
*
            USING RAND,15        SET UP ADDRESSABILITY FOR THIS SEGMENT
            ST    2,SAREA        STORE REG 2 IN SAVE AREA
            LM    1,2,KINDEX     LOAD INDEXES IN REGS 1 & 2
            LA    1,4(,1)        ADD 4 TO KINDEX
            C     1,=F'392'      IS KINDEX AT END OF TABLE
            BC    4,CKJX         IF NOT, CHECK JINDEX
            LA    1,0            SET KINDEX EQUAL TO 0
CKJX        LA    2,4(,2)        ADD 4 TO JINDEX
            C     2,=F'392'      IS JINDEX AT END OF TABLE
            BC    4,STINDX       IF NOT, THEN STORE INDEXES
            LA    2,0            SET JINDEX EQUAL TO 0
STINDX      STM   1,2,KINDEX     STORE NEW TABLE INDEXES
            L     0,M(1)         LOAD TABLE(K) INTO REG 0
            X     0,M(2)         EX-OR REG 0 WITH TABLE(J)
            ST    0,M(1)         STORE THE NEW TABLE(K)
            SLL   0,1            CHANGE TO FLOATING PT WORD
            ST    0,TEMP
            SDR   0,0            ZERO FLOATING PT REG 0
            AD    0,FLOAT        LOAD RAND IN FLOATING PT REG 0
            L     2,SAREA        RESTORE REG 2 TO OLD VALUE
            BR    14             RETURN TO CALLING PROGRAM
            DS    0D
FLOAT       DC    X'46000000'
TEMP        DS    F
KINDEX      DC    F'-4'
JINDEX      DC    F'104'
SAREA       DS    F
M           DC    X'14A37556'    TABLE
            DC    X'2343189D'
            DC    X'73D1D51D'
            DC    X'45EE222F'
            DC    X'69E891DE'
            DC    X'42B8F10A'
            DC    X'0A441D0C'
            DC    X'3AC17465'
            DC    X'5CA47405'
            DC    X'59980F0B'
            DC    X'26C447B3'
            DC    X'6076C207'
            DC    X'2624B1FB'
            DC    X'5652AFDB'
            DC    X'274A22F4'
            DC    X'3DAFD7CE'
            DC    X'10AD9CFB'
            DC    X'137C8FB4'
            DC    X'15E00410'
            DC    X'26A705B6'
```

```
DC    X'71D439D5'
DC    X'24DE8904'
DC    X'3265CB83'
DC    X'116B9245'
DC    X'11DBAF77'
DC    X'7F8B152C'
DC    X'3BF35749'
DC    X'17410F1C'
DC    X'4533E347'
DC    X'7941E1A0'
DC    X'13E982D7'
DC    X'21B6E123'
DC    X'07703D5D'
DC    X'7D6B1378'
DC    X'77CA6124'
DC    X'7730CFEF'
DC    X'04F5E1F4'
DC    X'5ED9C0C6'
DC    X'268BE30E'
DC    X'735B491D'
DC    X'2F62F010'
DC    X'288D6C36'
DC    X'699A2D62'
DC    X'4FF8E04E'
DC    X'6F737FB4'
DC    X'154381A2'
DC    X'3F4B9F96'
DC    X'24A76A42'
DC    X'0F9BEAED'
DC    X'29F60058'
DC    X'2F88558C'
DC    X'0A9FB8C6'
DC    X'61DA7AFA'
DC    X'5BC5E3C4'
DC    X'4F915DD2'
DC    X'3B0F9FA5'
DC    X'74AF50E6'
DC    X'6FF780E4'
DC    X'793AED2F'
DC    X'65FD67D5'
DC    X'7F75FBAF'
DC    X'269BA3DE'
DC    X'5CBE87D5'
DC    X'4820CA49'
DC    X'6AE1F384'
DC    X'6C78654A'

DC    X'1499C398'
DC    X'224C9BD6'
DC    X'61CF6FE9'
DC    X'5234CAA5'
DC    X'6D6F28DC'
DC    X'205D0157'
DC    X'4DD5C3E8'
DC    X'5804FC0D'
DC    X'576DC532'
DC    X'6DCEA885'
DC    X'6356E7D9'
DC    X'0CEF452A'
DC    X'28D7C142'
DC    X'35C4C46D'
DC    X'01EEE072'
DC    X'2F121062'
DC    X'29E3BF14'
DC    X'407074CB'
DC    X'5DB499A3'
DC    X'1329E4B0'
DC    X'2668F6BA'
DC    X'337F6F8C'
DC    X'5063400B'
DC    X'0507802B'
DC    X'749370E2'
DC    X'3B1DC32E'
DC    X'3AB12F7F'
DC    X'6AF175EF'
DC    X'6599C370'
DC    X'4ABD322F'
END
```

Appendix F:
Tausworthe Generator

```
RAND        CSECT
            USING RAND,15
            PRINT DOUBLE
*
*           TAUSWORTHE RANDOM NUMBER GENERATOR
*           USING THE PRIMITIVE POLYNOMIAL X**31 + X**13 + 1.
*           GENERATES A NORMALIZED RANDOM NUM. IN THE INTERVAL (0,1).
*           REFERENCES:     TAUSWORTHE,ROBERT C.,RANDOM NUMBERS GENERATED
*                   BY LINEAR RECURRENCE MODULO TWO,MATH. OF COMPUTATION,
*                   V. 19,1965, PP. 201-209,
*                           WHITTLESEY,F.R.B., COMPARISON OF THE CORREL-
*                   ATIONAL BEHAVIOR OF RANDOM NUMBER GENERATORS FOR THE IBM
*                   360,COMM OF THE ACM,V. 11,NUM. 9,SEPT 1968,PP. 641-644.
*           PROGRAMMER:  WARREN V. CAMP, U.S.L.
*
            L       0,SEED   LOAD REG 0 WITH PREVIOUS GENERATED RANDOM NUM.
            LR      1,0      COPY REG 0 INTO REG 1.
            SRL     1,13     RIGHT-SHIFT REG 1 13 PLACES.
            XR      1,0      EXCLUSIVE-OR REG 0 INTO REG 1.
            LR      0,1      STORE RESULT BACK INTO REG 0.
            SLL     1,18     LEFT-SHIFT REG 1 (31 - 13) PLACES.
            XR      0,1      EXCLUSIVE-OR REG 1 INTO REG 0.
            SLL     0,1      LEFT-SHIFT REG 0 1 PLACE.
            ST      0,SEED   STORE VALUE OF REG 0 IN SEED.
            SUR     0,0      ZERO OUT FLOATING PT REG 0.
            AD      0,FLOAT  LOAD NEW NORMALIZED RND NUM IN FLTING PT REG 0.
            SRL     0,1      RIGHT-SHIFT REG 0 1 PLACE.
            ST      0,SEED   STORE NEW RANDOM NUM. IN SEED.
            BR      14       RETURN TO CALLING PROGRAM.
            DS      0D       DEFINE CONSTANT AREA.
FLOAT       DC      X'46000000'   CHARACTERIST OF THE NEW RANDOM  NUM.
SEED        DC      F'1'     INITIALIZE SEED WITH THE VALUE OF 1.
            END
```

Appendixes for Tests

Appendix A:
Test Results

These are test results for five RNGs. An asterisk indicates failure at the 5 percent level of confidence. 10,000 numbers were used in each test.

Test		HD2	HD1	16 HD2	FSR	GEN FSR
Parameter		$a=14^{29}$	λ	16 *primitive* roots	$p=31$ $q=13$	$p=98;\ q=27$ 15 *bits/word*
Frequency: Real		102.2	106.4	100.0	89.5	95.3
123.2/59 Integer		87.8	94.3	92.1	102.7	111.2
Yules:	Digit 1	30.1	50.8	42.5	78.1*	40.9
	Digit 2	34.3	44.5	59.4	33.5	85.7*
63.0/	Digit 3	41.6	93.2*	42.9	40.0	30.8
23.3	Digit 4	39.2	36.3	41.8	63.2	41.2
	Digit 5	44.1	38.3	47.4	47.4	—
Gap: between	0	36.2	10.1	20.7	18.4	27.2
	1	20.3	20.7	18.6	11.7	23.2
	2	16.5	30.2	23.0	17.2	21.1
32.8/5.7	3	26.2	17.3	18.2	14.0	17.5
	4	20.1	20.7	20.6	14.4	19.8
	5	24.1	15.1	8.3	15.9	32.2
	6	31.5	37.5*	16.2	25.3	29.9
	7	19.1	7.1	13.6	27.2	16.1
	8	35.6*	16.5	20.1	20.9	27.4
Correlation: Lag		13	6	31	33	43
0.0, mean		0.0025	−0.0003	−0.0006	−0.0019	−0.0029
0.00010, var		0.00010	0.00009	0.00006	0.00011	0.00010
$0.03 \leq max \leq 0.08$		0.32	0.021*	0.020*	0.034	0.062
D^2: 106.3/47.8		91.0	84.2	98.9	62.3	50.7
Serial: 123.2/59.1		94.4	91.3	122.4	80.9	112.1
Runs: longest		6	6	6	8	6
6666±83, total		6642	6753*	6642	6690	6667
2500, above 0.5		2517	2525	2517	2539	2497
2500, below 0.5		2516	2525	2516	2539	2496
(11.1)/5 X^2 (14.1)/7		10.4/5	7.9/5	10.4/5	24.2/7*	2.3/5
Sum of $n =$	2	19.0*	64.2	25.4*	55.7*	50.9*
	3	46.9*	40.9*	73.7	15.6*	13.7*
123.2/591.	4	36.2*	17.8*	103.1	72.4	30.8*
	5	103.9	73.6	32.3*	65.8	10.5*
	10	25.9*	29.0	74.6	30.3*	29.7*
Min of	2	103.9	86.8	88.4	101.6	100.0
	4	108.3	99.2	111.7	110.6	103.2
	6	98.9	71.7	97.5	100.0	102.9

Test		HD2	HD1	16 HD2	FSR	GEN FSR
Parameter		$a=14^{29}$	λ	16 primitive roots	p=31 q=13	p=98; q=27 15 bits/word
	8	96.5	82.3	104.7	85.2	137.9*
	10	93.0	93.3	111.8	106.2	108.0
	12	71.0	95.6	99.5	109.8	123.2
	14	94.1	78.2	108.3	118.2	137.9*
	16	107.5	83.3	102.8	124.2*	120.6
	18	87.7	87.0	87.6	100.0	113.9
123.2/59.1	20	86.7	86.2	98.3	114.8	131.6*
Max of	2	101.8	108.8	105.4	137.2*	99.4
	4	75.5	101.5	118.4	117.0	88.1
	6	95.5	115.0	133.1*	95.4	93.5
	8	96.7	106.5	107.8	96.2	71.3
	10	106.1	98.2	109.6	106.4	101.3
	12	80.4	115.6	99.0	111.4	103.6
	14	82.2	100.8	127.7*	104.8	84.7
	16	83.1	91.2	105.6	109.0	99.8
	18	94.4	122.3	105.5	95.6	102.8
	20	100.1	114.0	96.9	124.4*	95.2

Conditional Bit — First twelve bits of each number

3.8/.00	2	0.68	2.79	0.4	0.54	2.35
7.8/.015	3	3.64	3.24	1.5	5.7	3.13
14.0/.49	4	16.7*	7.2	0.7	10.2	11.0
25.0/3.15	5	19.1	17.2	12.6	27.9*	16.3
45.0/115	6	28.9	37.5	22.0	23.7	21.6
82.5/32.8	7	54.0	63.6	53.6	70.0	47.2
154.9/80.0	8	114.9	119.8	100.9	127.0	129.3
306.5/185.0	9	239.5	245.4	252.7	272.0	237.0
562.1/405.0	10	497.7	532.7	491.9	568.0*	478.3
1099./860.	11	1035	1059	1056	1043	990.3
2155./900.	12	2089	2090	2077	2047	2088

Conditional Bit — Second twelve bits of each number

	2	0.67	0.40	1.2		0.62
0.80	3	3.62	2.42	4.8	0.87	5.64
	4	3.68	5.6	9.0	6.3	—
	5	14.0	13.2	20.3	18.2	—
	6	3ym5	32.7	25.6	38.6	—
	7	60.7	66.1	54.7	50.0	—
	8	120.0	136.5	119.1	154.0	—
	9	209.2	282.4	266.4	265.9	—
	10	486.7	535.7	490.9	539.6	—
	11	1022	943.9	1027	1026	—
	12	1957	2177	2190	2739	—

FFT: 1.65,	U	−0.531	−1.360	0.721	0.494	0.162
0.050≥	KS	0.025	0.008	0.048	0.032	0.041
2.35≤H≤67.5		149.6*	158.7*	166.8*	148.4*	138.6*
Scatter 48.6/ 10.2		36.8	30.7	53.4*	45.3	30.8

```
C      THIS PROGRAM IS USED TO CALL THE STATISTICAL TESTS USED TO
C      DETERMINE WHETHER A PARTICULAR RANDOM NUMBER GENERATOR IS
C      STATISTICALLY QUALIFIED.
C      PROGRAMMER:   WARREN V. CAMP, U.S.L.
       REAL RDATA(1300)
       LOGICAL TESTS(10),FINISH
       TESTS(1)=.TRUE.
       TESTS(2)=.TRUE.
       TESTS(3)=.TRUE.
       TESTS(4)=.TRUE.
       TESTS(5)=.TRUE.
       TESTS(6)=.TRUE.
       TESTS(7)=.TRUE.
       TESTS(8)=.TRUE.
       DO 3 I=1,8
       DO 1 J=1,1300
       RDATA(J)=RAND(NUM)
1      CONTINUE
2      IF(TESTS(1)) CALL FREQ(RDATA,FINISH)
       IF(TESTS(2)) CALL GAP(RDATA,FINISH)
       IF(TESTS(3)) CALL CNDBIT(RDATA,FINISH)
       IF(TESTS(4)) CALL MAXMIN(RDATA,FINISH)
       IF(TESTS(5)) CALL SERIAL(RDATA,FINISH)
       IF(TESTS(6)) CALL YULE(RDATA,FINISH)
       IF(TESTS(7)) CALL AUTOCR(RDATA,FINISH)
       IF(TESTS(8)) CALL DISTAN(RDATA,FINISH)
3      CONTINUE
       IF(FINISH) GO TO 999
       FINISH=.TRUE.
       GO TO 2
999    STOP
       END
```

129

Appendix C:
Frequency Test

```
      SUBROUTINE FREQ(RDATA,FINISH)
C
C     TEST:  THE FREQUENCY TEST.
C     PURPOSE:  DETECTS NONUNIFORMITY OF DIGITS.
C     PROCEDURE:  GENERATE NORMALIZED RANDOM NUMBERS.  ROUND THE NUMBERS
C     TO ONE DIGIT AND RECORD THEIR OCCURENCES IN A TABLE OF 10 EQUAL
C     SUBINTERVALS.  COMPARE THE EMPIRICAL DISTRIBUTION WITH THE
C     THEORETICAL WITH A CHI-SQUARE TEST.
C     REFERENCES:  KENDAL,M. G. AND SMITH, B. B.,RANDOMNESS AND RANDOM
C     SAMPLING NUMBERS, ROYAL STATISTICAL SOCIETY, LONDON JOURNAL, V.101,
C     1938, PG. 147-166.
C     PROGRAMMER:  WARREN V. CAMP, U.S.L.
C
      REAL RDATA(1300)
      LOGICAL FINISH
      INTEGER FREQAY(10)
      DATA FREQAY,NTIMES/11*0/
      IF(FINISH) GO TO 2
      DO 1 I=1,1300
      IX=RDATA(I)*10.0+1.0
      FREQAY(IX)=FREQAY(IX)+1
    1 CONTINUE
      NTIMES=NTIMES+1300
      RETURN
    2 WRITE(6,103) NTIMES
      WRITE(6,101) FREQAY
      FCHISQ=0.0
      TRUE=FLOAT(NTIMES)/10.0
      DO 3 J=1,10
      FCHISQ=FCHISQ+((FLOAT(FREQAY(J))-TRUE)**2)
    3 CONTINUE
      FCHISQ=FCHISQ/TRUE
      WRITE(6,104)
      WRITE(6,102) FCHISQ
      RETURN
  101 FORMAT('0COUNT IN EACH OF THE 10 CELLS:',//,' ',10(I6,6X))
  102 FORMAT('0RESULT OF THE CHI SQ TEST ON THE FREQ TEST IS',E14.7)
  103 FORMAT('1RESULTS OF THE FREQUENCY TEST USING ',I7,' SAMPLES FROM
     1RAND.')
  104 FORMAT('0FOR RAND TO PASS THE FREQ TEST THE RESULT OF THE CHI SQ T
     1EST, FOR A LEVEL OF SIGNIFICANCE = .05 AND DEGREE OF FREEDOM = 9,
     2',//,'    MUST BE EQUAL TO OR LESS THAN 16.9.')
      END
```

Appendix D:
Serial Test

```
      SUBROUTINE SERIAL(RDATA,FINISH)
C
C     TEST: THE SERIAL TEST.
C     PURPOSE:  DETECTS FOR NONUNIFORMITY IN PAIRS OF DIGITS.
C     PROCEDURE:  GENERATE A PAIR OF NORMALIZED RANDOM NUMBERS.  ROUND
C     THE NUMBERS TO ONE DIGIT AND RECORD THEIR OCCURENCES IN A TABLE
C     OF 100 EQUAL SUBINTERVALS.  THE EMPIRICAL DISTRIBUTION IS COMPARED
C     TO THE THEORETICAL BY A CHI-SQUARE TEST.
C     REFERENCES:  KENDAL,M. G. AND SMITH, B. B.,RANDOMNESS AND RANDOM
C     SAMPLING NUMBERS, ROYAL STATISTICAL SOCIETY, LONDON JOURNAL, V.101,
C     1938, PG. 147-166.
C     PROGRAMMER:  WARREN V. CAMP, U.S.L.
C
      REAL RDATA(1300)
      INTEGER SERARY(10,10)
      LOGICAL FINISH
      DATA SERARY/100*0/,NTIMES/0/,CHISQ/0.0/
      IF(FINISH) GO TO 2
      DO 1 I=1,1299,2
      J=RDATA(I)*10.0+1.0
      K=RDATA(I+1)*10.0+1.0
      SERARY(K,J)=SERARY(K,J)+1
    1 CONTINUE
      NTIMES=NTIMES+650
      RETURN
    2 WRITE(6,101) NTIMES
      WRITE(6,102) SERARY
      TRUE=FLOAT(NTIMES)/100.0
      DO 3 I=1,10
      DO 3 J=1,10
      CHISQ=CHISQ+((FLOAT(SERARY(I,J))-TRUE)**2)
    3 CONTINUE
      CHISQ=CHISQ/TRUE
      WRITE(6,103)
      WRITE(6,104) CHISQ
      RETURN
  101 FORMAT('1SERIAL TEST USING ',I6,' PAIRED SAMPLES FROM RAND')
  102 FORMAT('0COUNT FROM EACH OF THE 100 CELLS:',//,10('0',10(I5,5X),//
     1))
  103 FORMAT('0FOR RAND TO PASS THE SERIAL TEST THE RESULT OF THE CHI SQ
     1 TEST FOR A LEVEL OF SIGNIFICANCE = .05 AND DEGREE OF FREEDOM = 99
     2',//,'    MUST BE EQUAL TO OR LESS THAN 123.9.')
  104 FORMAT('0THE RESULT OF THE CHI SQ TEST IS ',E14.7)
      END
```

Appendix E:
Gap Test

```
      SUBROUTINE GAP(RDATA,FINISH)
C
C     TEST:  THE GAP TEST.
C     PURPOSE:  DETECTS UNEXPECTED DISTRIBUTIONS IN REGARD TO THE
C     GAP OCCURING BETWEEN THE SAME DIGITS IN THE SERIES.
C     PROCEDURE:  RECORD THE OCCURENCES OF THE LENGTH OF GAPS BETWEEN
C     SUCCESSIVE LIKE DIGITS.  COMPARE THE EMPIRICAL DISTRIBUTION WITH
C     THE THEORETICAL GEOMETRIC DISTRIBUTION.
C     REFERENCES:  KENDAL,M. G. AND SMITH, B. B.,RANDOMNESS AND RANDOM
C     SAMPLING NUMBERS, ROYAL STATISTICAL SOCIETY, LONDON JOURNAL, V.101,
C     1938, PG. 147-166.
C     PROGRAMMER:  WARREN V. CAMP, U.S.L.
C
      REAL RDATA(1300)
      INTEGER GAPARY(22),GCTR
      LOGICAL FINISH
      DATA GAPARY,GCTR,NTIMES,IGAP,PTOT,GCHISQ/24*0,1,2*0,0/
      IF(FINISH) GO TO 3
      DO 2 I=1,1300
      IX=RDATA(I)*10.0
      IF(IX.EQ.2) GO TO 1
      IGAP=IGAP+1
      GO TO 2
1     IF(IGAP.GT.22) IGAP=22
      GAPARY(IGAP)=GAPARY(IGAP)+1
      IGAP=1
      GCTR=GCTR+1
2     CONTINUE
      NTIMES=NTIMES+1300
      RETURN
3     WRITE(6,101) NTIMES,GCTR
      WRITE(6,102) GAPARY
      DO 7 J=1,22
      IF(J.NE.22) GO TO 5
      PL=1.0-PTOT
      GO TO 6
5     PL=(.1)*((.9)**(J-1))
      PTOT=PTOT+PL
6     PLN=PL*FLOAT(GCTR)
      GCHISQ=GCHISQ+(((FLOAT(GAPARY(J))-PLN)**2)/PLN)
7     CONTINUE
      WRITE(6,103)
      WRITE(6,104) GCHISQ
      RETURN
101   FORMAT('1RESULTS OF THE GAP TEST USING ',I7,' SAMPLES FROM RAND PR
     1ODUCING ',I5,' GAPS FOR THE DIGIT 2.')
102   FORMAT('0COUNT IN EACH OF THE 22 CELLS:',2(//,' ',11(I6,4X)))
103   FORMAT('0FOR RAND TO PASS THE GAP TEST THE RESULT OF THE CHI SQ TE
     1ST, FOR A LEVEL OF SIGNIFICANCE = .05 AND DEGREE OF FREEDOM 21,',/
     2/,'    MUST BE EQUAL TO OR LESS THAN 32.7.')
104   FORMAT('0RESULT OF THE CHI SQ TEST IS = ',E14.7)
      END
```

Appendix F:
Yule Test

```
      SUBROUTINE YULE(RDATA,FINISH)
C
C     TEST:  THE YULE TEST.
C     PURPOSE:  TO DETECT FOR PATCHINESS.
C     METHOD:  TAKE THE SUM OF FIVE RANDOM DIGITS AND COMPARE THE
C     DISTRIBUTION OF THE SUMS WITH THE EXPECTED.
C     REFERENCES:  YULE,G.UDNY,ROYAL STATISTICAL SOCIETY, LONDON JOURNAL
C     V.101,1938,PG.167-172.
C     PROGRAMMER:  WARREN V. CAMP, U.S.L.
C
      REAL RDATA(1300),PROB(46)
      INTEGER YULARY(46)
      LOGICAL FINISH
      DATA NTIMES,YULARY/47*0/,CHISQ/0.0/
      DATA PROB/.00001,.0000499,.00015,.0003499,.0006999,.00126,.0021,.0
     103299,.0049499,.0071499,.0099599,.0134,.01745,.02205,.0271,.03246,
     2.03795,.0433499,.0484,.052799,.05631,.05875,.06,.06,.05875,.05631,
     3.052799,.0484,.0433499,.03795,.03246,.0271,.02205,.01745,.0134,.00
     499599,.0071499,.0049499,.003299,.0021,.00126,.000699,.0003499,.000
     515,.00004999,.00001/
      IF(FINISH) GO TO 3
      DO 2 I=1,1300,5
      ISUM=1
      JSTART=I
      JEND=I+4
      DO 1 J=JSTART,JEND
      IELEM=RDATA(J)*10.0
      ISUM=ISUM+IELEM
    1 CONTINUE
      YULARY(ISUM)=YULARY(ISUM)+1
    2 CONTINUE
      NTIMES=NTIMES+260
      RETURN
    3 WRITE(6,101) NTIMES
      WRITE(6,102) YULARY
      DO 4 I=1,46
      TRUE=FLOAT(NTIMES)*PROB(I)
      CHISQ=CHISQ+((FLOAT(YULARY(I))-TRUE)**2)/TRUE
    4 CONTINUE
      WRITE(6,103)
      WRITE(6,104) CHISQ
      RETURN
  101 FORMAT('1YULE TEST USING ',I6,' SUMS OF 5 DIGITS FROM RAND')
  102 FORMAT('0COUNT FROM EACH OF THE 46 CELLS:',//,23('0',2(I6,4X),//))
  103 FORMAT('0FOR RAND TO PASS THE YULE TEST THE RESULT OF THE CHI-SQ T
     1EST FOR A LEVEL OF SIGNIFICANCE= .05 AND A DEGREE OF FREEDOM = 45,
     2',//,'  MUST BE EQUAL OR LESS THAN 61.8.')
  104 FORMAT('0RESULT OF THE CHI SQ TEST IS ',E14.7)
      END
```

Appendix G:
D² Test

```
      SUBROUTINE DISTAN(RDATA,FINISH)
C
C     TEST:  THE DISTANCE SQUARED TEST.
C     PURPOSE:  TO TEST DIGITS FOR LOCAL RANDOMNESS, WHEN THE DIGITS
C     ARE TO BE USED IN A MONTE CARLO METHOD.
C     PROCEDURE:  RANDOM DIGITS ARE USED TO SELECT A RANDOM PT IN THE
C     UNIT SQUARE, THE DIGITS THUS REPRESENT THE COORDINATES OF
C     A POINT BETWEEN (0,0) AND (1,1).  THE SQUARE OF THE DISTANCE IS
C     COMPUTED AND ROUNDED TO ONE DECIMAL PLACE.  THE DISTRIBUTION OF
C     THE RESULTS IS COMPARED TO THE THEORETICAL BY THE CHI-SQUARE TEST.
C     REFERENCES:  GRUENBEGER,FRED AND MARK,A.M.,MATH. TABLES AND
C     OTHER AIDS TO COMP.,V.5,1951,PG 109.
C     PROGRAMMER:  WARREN V. CAMP, U.S.L.
C
      REAL RDATA(1300),PROB(20)
      INTEGER DISTAY(20)
      LOGICAL FINISH
      DATA DISTAY,NTIMES/21*0/,CHISQ/0.0/,EMPIR/0.0/
      DATA PROB/.234832,.174973,.139495,.112718,.090969,.072614,.056748,
     1.042814,.030430,.019333,.010777,.006345,.003740,.002138,.001154,.0
     200572,.000246,.000084,.000017,.000001/
      IF(FINISH) GO TO 2
      DO 1 I=1,1300,4
      D=((RDATA(I)-RDATA(I+1))**2)+((RDATA(I+2)-RDATA(I+3))**2)
      ID=D*10.0+1.0
      DISTAY(ID)=DISTAY(ID)+1
1     CONTINUE
      NTIMES=NTIMES+325
      RETURN
2     WRITE(6,101) NTIMES
      WRITE(6,102) DISTAY
      DO 5 I=1,20
      D=FLOAT(I)/10.0
      TRUE=FLOAT(NTIMES)*PROB(I)
      CHISQ=CHISQ+((FLOAT(DISTAY(I))-TRUE)**2)/TRUE
5     CONTINUE
      WRITE(6,103)
      WRITE(6,104) CHISQ
      RETURN
101   FORMAT('1RESULTS OF THE DISTANCE TEST USING',I5,' DISTANCES SQUARE
     1D WITH VALUES FROM 0.0 TO 2.0 IN INCREMENTS OF .1')
102   FORMAT('0COUNT FROM EACH OF THE 20 CELLS:',//,2('0',10(I6,4X),//))
103   FORMAT('0FOR RAND TO PASS THE DISTANCE SQUARED TEST THE RESULT OF
     1THE CHI-SQ TEST, FOR A LEVEL OF SIGNIFICANCE=.05 AND A DEGREE OF
     2',//,' FREEDOM = 19, MUST BE EQUAL TO OR LESS THAN 30.1.')
104   FORMAT('0THE RESULT OF THE CHI SQ TEST IS = ',E14.7)
      END
```

Appendix H:
Autocorrelation Test

```
      SUBROUTINE AUTOCR(RDATA,FINISH)
C
C     TEST:  THE AUTOCORRELATION TEST.
C     PURPOSE:  TO INVESTIGATE CERTAIN EMPIRICAL PROPERTIES OF
C     AUTOCORRELATION FUNCTIONS CALCULATED FROM SEQUENCES OF NUMBERS.
C     PROCEDURE:  GENERATE STRINGS OF O MEAN RANDOM NUMBERS.  THE LENGTH
C     OF THE STRINGS SHOULD BE > OR = 1250.  CALCULATE THE NORMALIZED
C     CORRELATION FUNCTIONS WITH LAGS FROM 1 TO 50.  CONSTRUCT A TABLE
C     OF 50 EQUAL SUBINTERVALS AND RECORD THE MAXIMUM VALUE FOR EACH OF
C     THE CORRELATION FUNCTIONS.  COMPARE THE EMPIRICAL VALUES WITH
C     THE THEORETICAL.
C     REFERENCES:  WHITTLESEY,J.R.B.,A COMPARSION OF THE CORRELATIONAL
C     BEHAVIOR OF RANDOM NUMBER GENERATORS FOR THE IBM 360, COMM OF THE
C     ACM,V.11,NUM.9,SEPT 1968,PG. 641-644.
C     PROGRAMMER:  WARREN V. CAMP, U.S.L.
C
      DIMENSION ACORR(1300),R(51),RMAX(50)
      REAL RDATA(1300)
      LOGICAL FINISH
      DATA RMAX/50*0.0/,NTIMES/0/
      IF(FINISH) GO TO 40
      SEQLEN=1250.0
      ITIMES=SEQLEN
      MTIMES=ITIMES+50
10    DO 1 N=1,MTIMES
      ACORR(N)=RDATA(N)-0.5
1     CONTINUE
      DO 20 J=1,51
      R(J)=0.0
      K=J-1
      DO 25 I=1,ITIMES
      R(J)=R(J)+ACORR(I)*ACORR(I+K)
25    CONTINUE
      R(J)=R(J)/SEQLEN
      IF(J.EQ.1) GO TO 20
      R(J)=R(J)/R(1)
      IF(RMAX(J-1).LT.ABS(R(J))) RMAX(J-1)=ABS(R(J))
20    CONTINUE
      NTIMES=NTIMES+1300
      RETURN
40    WRITE(6,102) NTIMES
      WRITE(6,103)
      WRITE(6,101) RMAX
      RETURN
101   FORMAT(10('0',5(F14.7,5X),//))
102   FORMAT('1RESULTS OF THE AUTOCORRELATION TEST USING ',I5,' SAMPLES
     1FROM RAND.')
103   FORMAT('0FOR RAND TO PASS THIS TEST EACH OF THE 50 CELLS MUST HAVE
     1 A VALUE BETWEEN .03 AND .08.')
      END
```

Appendix I:
Max-min Test

```
      SUBROUTINE MAXMIN(RDATA,FINISH)
C
C     TEST:  THE MAXMIN TEST.
C     PURPOSE:  TO CHECK FOR NONUNIFORMITY OF THE MAXIMUM AND MINIMUM
C     VALUES.
C     METHOD:  SELECT THE LARGEST(SMALLEST) VALUE OF A SET OF N NUMBERS.
C     RAISE THIS VALUE(1-VALUE) TO A POWER OF N. A CHI-SQ TEST IS PER-
C     FORMED ON 100 EQUAL SUBINTERVALS.
C     REFERENCES:  MACLAREN,M.D. AND MARSAGLIA,G., UNIFORM RANDOM NUMBER
C     GENERATORS, J.ACM 12,1965,PG. 87-88.
C     PROGRAMMER:  WARREN V. CAMP, U.S.L.
C
      REAL RDATA(1300)
      INTEGER MAXARY(100),MINARY(100)
      LOGICAL FINISH
      DATA NTIMES,MAXARY,MINARY,CHISQ/201*0,0.0/
      IF(FINISH) GO TO 3
      DO 2 I=1,1300,5
      JSTART=I
      JEND=I+4
      RMIN=1.0
      RMAX=0.0
      DO 1 J=JSTART,JEND
      IF(RDATA(J).LT.RMIN) RMIN=RDATA(J)
      IF(RDATA(J).GT.RMAX) RMAX=RDATA(J)
1     CONTINUE
      IMAX=RMAX**5*100.0+1.0
      MAXARY(IMAX)=MAXARY(IMAX)+1
      IMIN=(1.0-RMIN)**5*100.0+1.0
      MINARY(IMIN)=MINARY(IMIN)+1
2     CONTINUE
      NTIMES=NTIMES+260
      RETURN
3     WRITE(6,101) NTIMES
      WRITE(6,102) MAXARY
      TRUE=FLOAT(NTIMES)/100.0
      DO 4 I=1,100
      CHISQ=CHISQ+((FLOAT(MAXARY(I))-TRUE)**2)
4     CONTINUE
      CHISQ=CHISQ/TRUE
      WRITE(6,103)
      WRITE(6,104) CHISQ
      WRITE(6,105) NTIMES
      WRITE(6,102) MINARY
      CHISQ=0.0
      DO 6 I=1,100
      CHISQ=CHISQ+((FLOAT(MINARY(I))-TRUE)**2)
6     CONTINUE
      CHISQ=CHISQ/TRUE
      WRITE(6,103)
      WRITE(6,104) CHISQ
      RETURN
1SETS OF 5 RANDOM NUMBERS.')
102   FORMAT('0COUNT IN EACH OF THE 100 CELLS:',//,10('0',10(I6,4X),//))
103   FORMAT('0FOR RAND TO PASS THE MAXMIN TEST THE RESULT OF THE CHI SQ
     1 TEST FOR A LEVEL OF SIGNIFICANCE = .05 AND DEGREE OF FREEDOM = 99
     2,',//,'    MUST BE EQUAL TO OR LESS THAN 123.2.')
104   FORMAT('0THE RESULT OF THE CHI SQ TEST IS ',E14.7)
105   FORMAT('1RESULTS OF THE MAXMIN TEST FOR',I6,' MINIMUM VALUES FROM
     1SETS OF 5 RANDOM NUMBERS.')
      END
```

143

```
      SUBROUTINE CNDBIT(RDATA,FINISH)
C
C
C     TEST:  THE CONDITIONAL BIT TEST.
C     PURPOSE:  TEST FOR DEPENDENCE OF BITS IN A RANDOM NUMBER.
C     METHOD:  CONSTRUCT A TABLE OF ALL POSSIBLE BIT CONFIGURATIONS
C     OF THE SEQUENCE OF BITS UNDER CONSIDERATION.  PERFORM A CHI-SQ
C     TEST ON THE EMPIRICAL COUNTS.
C     REFERENCES:  LEWIS, T. G., COMPUTER PROGRAMMING FOR PROBABILITY
C     DISTRIBUTION SAMPLING, UNPUBLISHED REPORT.
C     PROGRAMMER:  WARREN V. CAMP, U.S.L.
C
      REAL RDATA(1300),CHIARY(4)
      INTEGER CBITAY(4,8),IDF(4)
      LOGICAL LBYTE*1(4),LWORD*1(4),FINISH
      EQUIVALENCE (LBYTE(1),RBYTE),(LWORD(1),IWORD)
      DATA NTIMES,CBITAY/33*0/
      DATA CHIARY,IDF/3.84,3.84,7.81,14.1,1,1,3,7/
      IF(FINISH) GO TO 26
      IWORD=0
      DO 21 I=1,1300
      LWORD(4)=LBYTE(2)
      IF(IWORD.NE.64) GO TO 20
      LWORD(4)=LBYTE(1)
      RBYTE=RDATA(I)
      IWORD=IWORD/16
      GO TO (1,2,3,4,5,6,7,8,9,10,11,12,13,14,15),IWORD
    1 CBITAY(4,1)=CBITAY(4,1)+1
      GO TO 20
    2 CBITAY(3,1)=CBITAY(3,1)+1
      GO TO 20
    3 CBITAY(3,1)=CBITAY(3,1)+1
      CBITAY(4,2)=CBITAY(4,2)+1
      GO TO 20
    4 CBITAY(2,1)=CBITAY(2,1)+1
      GO TO 20
    5 CBITAY(2,1)=CBITAY(2,1)+1
      CBITAY(4,3)=CBITAY(4,3)+1
      GO TO 20
    6 CBITAY(2,1)=CBITAY(2,1)+1
      CBITAY(3,2)=CBITAY(3,2)+1
      GO TO 20
    7 CBITAY(2,1)=CBITAY(2,1)+1
      CBITAY(3,2)=CBITAY(3,2)+1
      CBITAY(4,4)=CBITAY(4,4)+1
      GO TO 20
    8 CBITAY(1,1)=CBITAY(1,1)+1
      GO TO 20
    9 CBITAY(1,1)=CBITAY(1,1)+1
      CBITAY(4,5)=CBITAY(4,5)+1
      GO TO 20
   10 CBITAY(1,1)=CBITAY(1,1)+1
      CBITAY(3,3)=CBITAY(3,3)+1
   11 CBITAY(1,1)=CBITAY(1,1)+1
      CBITAY(3,3)=CBITAY(3,3)+1
      CBITAY(4,6)=CBITAY(4,6)+1
      GO TO 20
   12 CBITAY(1,1)=CBITAY(1,1)+1
      CBITAY(2,2)=CBITAY(2,2)+1
      GO TO 20
```

```
13      CBITAY(1,1)=CBITAY(1,1)+1
        CBITAY(2,2)=CBITAY(2,2)+1
        CBITAY(4,7)=CBITAY(4,7)+1
        GO TO 20
14      CBITAY(1,1)=CBITAY(1,1)+1
        CBITAY(2,2)=CBITAY(2,2)+1
        CBITAY(3,4)=CBITAY(3,4)+1
        GO TO 20
15      CBITAY(1,1)=CBITAY(1,1)+1
        CBITAY(2,2)=CBITAY(2,2)+1
        CBITAY(3,4)=CBITAY(3,4)+1
        CBITAY(4,8)=CBITAY(4,8)+1
20      NTIMES=NTIMES+1
21      CONTINUE
        RETURN
26      CBITAY(1,2)=FLOAT(NTIMES)-CBITAY(1,1)
        DO 24 I=1,4
        CHISQ=0.0
        NCELLS=2**(I-1)
        TRUE=FLOAT(NTIMES)/2.0**I
        IF(I.NE.1) GO TO 28
        JEND=2
        GO TO 30
28      JEND=2**(I-1)
30      DO 22 J=1,JEND
        CHISQ=CHISQ+((FLOAT(CBITAY(I,J))-TRUE)**2)
22      CONTINUE
        CHISQ=CHISQ/TRUE
        WRITE(6,101) I,NTIMES
        WRITE(6,102) NCELLS,(CBITAY(I,L),L=1,NCELLS)
        WRITE(6,103) IDF(I),CHIARY(I)
        WRITE(6,104) CHISQ
24      CONTINUE
        RETURN
101     FORMAT('1RESULTS OF THE CONDITIONAL BIT TEST PERFORMED BY TESTING
       1THE ',I2,'TH BIT OF EVERY RANDOM NUM USING',I7,' SAMPLES FROM RAND
       2.')
102     FORMAT('0COUNT IN EACH OF THE ',I2,' CELLS:',//,' ',8(I6,6X))
103     FORMAT('0FOR RAND TO PASS THE COND BIT TEST THE RESULT OF THE CHI
       1SQ TEST , FOR LEVEL OF SIGNIFICANCE = .05 AND DEGREE OF FREEDOM ='
       2,I3,', ',//,'   MUST BE EQUAL TO OR LESS THAN ',F5.2,'.')
104     FORMAT('0RESULT OF THE CHI SQ TEST IS ',E14.7)
        END
```

Index

Index

About the Author

T. G. Lewis is an associate professor of computer science at the University of Southwestern Louisiana. He received the Ph.D. in computer science at Washington State University. Formerly, he was a programmer with Boeing Company and an engineer with the Sylvania Defense Lab. Dr. Lewis has published "Conditional Bit Sampling: Accuracy and Speed," with W. H. Payne in *Mathematical Software,* Academic Press, 1971, and "Generalized Feedback Shift Register Algorithm," with W. H. Payne in *Journal of the ACM*. His other activities include serving as director of SICMINI, a mini-computer special interest committee within ACM and a director of the Microdata Users Group.